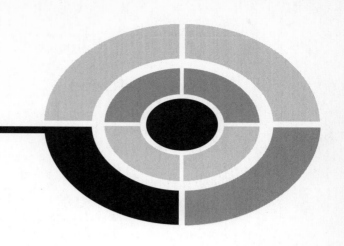

Robotics
Demystified

Demystified Series

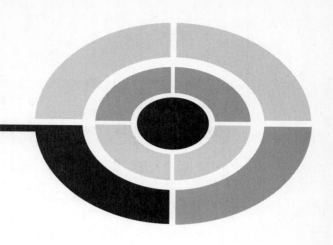

Robotics
Demystified

EDWIN WISE

McGRAW-HILL
New York   Chicago   San Francisco   Lisbon   London
Madrid   Mexico City   Milan   New Delhi   San Juan
Seoul   Singapore   Sydney   Toronto

I dedicate this book to all of the brain cells I lost writing it, and my amazing wife Marla who, against all good sense, lets me write books.

The McGraw·Hill Companies

Cataloging-in-Publication Data is on the file with the Library of Congress

4 5 6 7 8 9 0   DOC/DOC   0 1 0 9 8

ISBN 0-07-143678-2

The sponsoring editor for this book was Judy Bass and the production supervisor was Pamela A. Pelton. It was set in Times Roman by Keyword Publishing Services Ltd. The art director for the cover was Margaret Webster-Shapiro; the cover designer was Handel Low.

Printed and bound by RR Donnelley.

 This book is printed on recycled, acid-free paper containing a minimum of 50% recycled, de-inked fiber.

McGraw-Hill books are available at special quantity discounts to use as premiums and sales promotions, or for use in corporate training programs. For more information, please write to the Director of Special Sales, McGraw-Hill Professional, Two Penn Plaza, New York, NY 10121-2298. Or contact your local bookstore.

CONTENTS

CONTENTS

CONTENTS

PREFACE

This book is for people who want to learn the basic concepts of robotics without taking a formal course. This book seeks to give you an intuitive grasp of the various technologies that make up the field of robotics. There is no one "robot technology," so this book breaks the study of robots down into technology categories: the mechanics and framework of the robot, the electronics that make up its brain and nerves, and the control systems and programming that gives the robot life.

To aid you in your learning, this work contains short quizzes that review the main concepts in each chapter. Answers to the quizzes are given in the back of the book.

This is a hands-on book, and it contains a number of experiments to illustrate the concepts described in the text. These experiements can be completed with easily found materials.

Though this is a hands-on book, it is not a "how to" book. Since it covers several complex topics, it can only go so deep into each one.

Edwin Wise

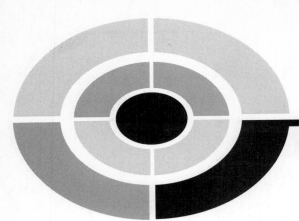

ACKNOWLEDGMENTS

Of course, I have to acknowledge my wife Marla for letting me take the time to write this book. My reviewers, Bob Comer and Matt Pinsonneault with help from Nikolas Wise, helped me to remain coherent and stay on track.

The text was written in Microsoft Word, and the graphics and diagrams were created and edited with Adobe's Photoshop, Midnight Software's DeltaCAD, CadSoft's Eagle electronics layout editor, Pacestar's ever useful EDGE Diagrammer, and MFSoft's Equation Grapher. Facts were checked against the Oxford English Dictionary and the Wikipedia at www.wikipedia.com, both of which are excellent sources of information.

Special appreciation goes out to the fine people at www.lugnet.com and the tools their members have created, without which this book would look a lot less interesting. MLCad by Michael Lachmann let me draw all of the LEGO designs and LPub by Kevin Clague processed these models, using geometry in Luts Uhlmann's LGeo library, for rendering.

Electronic layout to POVRay model conversion was handled by Eagle3D, which was written by Matthias Weisser. Eagle3D can be found on his homepage at http://www.matwei.de/. All rendering was done by POVRay, from www.povray.org.

LEGO is a trademark of the LEGO Group of companies, who reserve all rights to their names and logos. This is not a LEGO book, a book about LEGOs, nor is this work endorsed by LEGO. We do, however, use the LEGO construction system to provide mechanical examples. These little plastic blocks are available everywhere, and provide a familiar context for our exploration. The LEGO Mindstorms system also provides motors and computers that we use to learn about robotic systems.

All other product names are the trademark of their respective companies.

E. W.

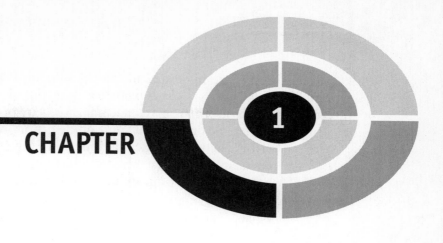

CHAPTER 1

Introduction

Robot

1.a. One of the mechanical men and women in Čapek's play; hence, a machine (sometimes resembling a human being in appearance) designed to function in place of a living agent, esp. one which carries out a variety of tasks automatically or with a minimum of external impulse.

  b. A person whose work or activities are entirely mechanical; an automaton.

Oxford English Dictionary, Online Edition

Karel Čapek used the word *Robot* in his 1921 play *Rossum's Universal Robots*, derived from the Czech word *robota*, meaning "forced labor." These Robots were created to replace man and, in their simplified form, as cheap labor.

Robots had perfect memory but were incapable of thinking new thoughts. They mirrored the Hebrew legends of the *golem*, a clay statue that has had life breathed into to by mystical means. And, of course, this all sounds a lot like Dr. Frankenstein's monster, reanimated from the bits and pieces dug up from the local graveyard.

One thing these stories have in common is that the creation is ultimately the downfall of their creator—robots, golems, and reanimated flesh mean *trouble*. They are an illustration of what happens when we reach too far and are bitten by the unintended consequences.

We, however, are interested in the robot as an *agent that carries out its tasks automatically or with a minimum of external impulse* rather than a recreation of life itself. A smart machine.

In this chapter, we first look at some of the history behind the robot. From there we explore the technologies that make up a robot, laying the groundwork for later chapters on how these technologies work. Once we have a good sense of what a robot is, we peek into the future to see what robots might someday be like.

A Brief Tour of Robotics

AUTOMATA AND ANIMATRONICS

An automaton is a device that has the ability to move under its own power. The mechanism of the motion is normally hidden, giving the illusion that the device is self-motivated or alive. While this definition can apply to something as mundane as a mechanical watch, automata are usually mechanisms that try to mimic the look and behavior of living creatures.

We humans have long been fascinated by the workings of our own bodies and the animals around us. With this fascination has come the urge to *recreate* these things, to step into the role of divinity and try our hand at the game of life.

The ancient Greeks, at around 400 B.C. and continuing on into the common era, are reputed to have used steam and water power to animate statues or drive various mechanisms in their temples. Automatically opening doors, statues that appear to drink offerings of wine, singing birds, self-lighting fires, and other wonders are documented in the few remaining writings of that time. There are hints of similar Egyptian and Chinese devices from that era as well.

Most of these technologies, the accumulated knowledge of ancient civilizations, were lost until relatively recent times. During the Renaissance, Europe started to drag itself out of the Dark Ages and began discovering (or, in many cases, rediscovering) all manner of ideas, art, technologies, and sciences. Among these, combining both art and technology, were the automata.

Some wild stories tell us about an iron fly and an artificial eagle made of wood, constructed by Johannes Muller in the 1470s. In the fourteenth and fifteenth centuries, automata were the playthings of royalty. Leonardo da Vinci made an animated lion for King Louis XII, Gianello della Tour of Cremona built a number of mechanical entertainers for Emperor Charles V, and Christiaan Huygens created a robotic army sometime around 1680.

The first documented automaton in human form, or android, was made by Hans Bullman in the early sixteenth century. Androids have been a popular subject for automata builders ever since. Inventors built machines to play musical instruments of all kinds, draw, write, and even play chess—or at least, pretend to play chess.

The eighteenth century was the golden age of automata, with many intricate machines. These were driven by clockwork gears and cylinders containing hundreds, if not thousands, of complicated control tracks. These tracks were composed of sequences of rods of different heights fixed to a cylinder, or individual cams with complex shapes, that pushed on levers that moved rods that adjusted the automaton creating a specific sequence of actions.

The Turk was a world-famous automaton from this time. Built in 1770 by Wolfgang von Kepelen, and later purchased from Kepelen's son by Johan Nepomuk Maelzel in 1804, the Turk toured Europe and America amazing audiences by playing chess!

By the time the Turk was on tour, audiences were familiar with the workings of automata and had been exposed to many fine machines. But they were also confident that these machines were just that, simple collections of gears and levers whose rote actions were no challenge to the human intellect. The automata may appear to be alive, but they are only vague shadows of life. They couldn't think.

The Turk challenged this view. It played, and often won, games of chess against any number of famous figures of the time. Napolean, Charles Babbage, and Edgar Allen Poe all took their turn against this mechanical savant. Of course, it turned out that the machine could not play chess at all. Instead, it provided cramped quarters for a human chess player who in turn ran the machinery that made the Turk move.

One very complex automaton wasn't an android, but a duck. Jacques Vaucanson created this avian automaton in 1738 and then went on tour with it. At the price of a week's wages, audiences were invited to see this creation move around, adjust its wings, preen, drink water, and even eat food, digest it, and then defecate. All of this required thousands of moving parts within both the duck and its large base. And yet, automata were just a hobby of Vaucanson's. He sold his collection in 1743 and went on to direct the state-owned silk-mills in France. Among other innovations, he developed a way to weave silk brocade using a machine guided by perforated cards. Owing to hostility among the weavers of the time, his advances in factory automation were ignored for decades.

In 1804, Joseph-Marie Jacquard improved and reintroduced the technique and was later credited with its invention. While the automatic loom was still despised by weavers, who went as far as burning down automated factories,

its improved efficiency led to its ultimate acceptance and led the way into the industrial revolution.

In the nineteenth century improved manufacturing techniques brought simple automata to the masses, typically in the form of toys, fancy clocks, and other novelties. Clockwork mechanical toys were popular well into the twentieth century. Today the springs, gears, and cams in toys have been replaced by tiny motors and electronic controls.

The skills and techniques developed by the automata makers during the Renaissance provided a foundation for the industrial revolution that followed. Today, you can still find automata for sale. Automata are now in the domain of the artist and pieces from modern craftsmen and artists can be found for as little as a few dollars, up to hundreds or thousands of dollars.

Another use for these magical machines is entertainment. Walt Disney introduced mechanical actors in the displays of his amusement park and christened them Audio-Animatronics. This cumbersome name is normally shortened to simply "animatronics." Animatronics are machines driven by motors and hydraulics and synchronized with an audio track to give the full illusion of life.

From the simplest Egyptian trick with water to the modern miracles of Disney's animatronics, these machine all share one characteristic. They can only reproduce a preset sequence of motions.

FACTORY MACHINES

Ever since the advent of factories during the industrial revolution, specialized machines have had an important role in creating the products of civilization. The most common machine was the underpaid, overworked citizen—men, women, and children. Early factory conditions were dangerous, but the wages were good and nobody could argue with the efficiency factories brought.

Water and steam power, and later gas and electric power, replaced and enhanced human power, allowing us to make our products even faster and cheaper. Complex machines were created to take over many aspects of manufacture. The automatic loom is well known, but even today there are specific machines for many tasks.

You don't normally think about it, but there is a complex machine whose only purpose is to bend wire into paperclips. There is another machine, perhaps in the same factory, that makes nails. Other machines perform other tasks. These machines, invaluable as they are for industry, are still forms of automata.

Factory automata start to become robots when they gain the ability to be programmed. But there is still a large gray area. Take that nail-making machine and add a bunch of controls to it so it can make nails from different sizes of wires, with different types of points, and different types of heads. Is it an automaton or a robot? Does it make a difference if the controls are mechanical levers and knobs or electronic circuits?

In the early factories, working alongside a machine made your job more dangerous even if it made it less arduous. These early machines were large assemblies of spinning, whirring, moving parts that continued to spin, whir, and move even if a finger, foot, or other body part intruded into it. Even today, people working with machines in factories and food-processing plants face special risks. Machines are designed to be as safe as possible, but there are limits to what can be done to a metal sheer or punch press, for example, and have it remain useful.

As machines improved into robots, they made some aspects of factory work safer. A robotic painter, spot welder, or assembly machine can operate in an empty space without any help at all. A supervisor stands safely outside its range of motion while the robot does the dirty and dangerous work (Fig. 1-1).

The most visible type of factory robot is the robot arm (Fig. 1-2). These can be given any type of specialized "hand" needed for their job (Fig. 1-3) and programmed to perform complex activities. One arm, with a set of different hands, can be programmed to perform any number of tasks. These are the robots that we recognize as "smart" machines, beginning to realize the dream promised to us by Kepelen's Turk.

Fig. 1-1.   Welding robot (photo courtesy Motoman).

Fig. 1-2.   Arm with welding attachment (photo courtesy Motoman).

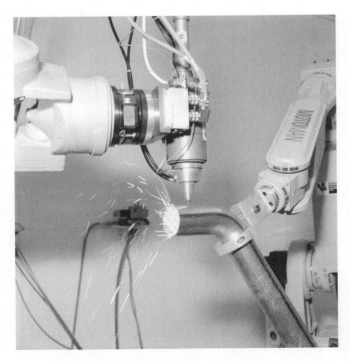

Fig. 1-3.   Cutting attachment (photo courtesy Motoman).

Robots make some exploration jobs not only safer but *possible*. Most humans would not be able to walk into the mouth of an active volcano, perform hazardous-waste cleanup at the site of a nuclear accident, explore the surface of Mars for months on end, or crawl through the debris of a fallen building looking for survivors.

FICTIONAL ROBOTS

Though not robots, legends tell about the creation of artificial life through many methods, most of which are poorly defined. How can we repeat the feats of the golem makers and the Greek deities who breathed life into clay? Even the first use of the word robot by Karel Čapek was referring to a creature that was more biological than mechanical, a precursor to Mary Shelley's Frankenstein monster.

Real robots are mechanical and reproducible, machines that are built following clever and complex blueprints and driven by ingenious programs. But even in the world of mechanical men, the vision of what a robot *could* be has always raced ahead of what we can actually build.

C3PO and R2D2 from *Star Wars* were the robots of my generation's dreams. More recently, expectations were raised with the cyborgs of *Blade Runner* and even more so with the T2000 "liquid metal" robot from *Terminator 2* and *3*.

Going back in time, there were robots spanning the spectrum from the androids of *West World*, the flailing robot from the original *Lost in Space* television series (borrowed from the even older movie *Forbidden Planet*), or even the stumpy repair robots from *Silent Running*.

But why stop in the 1950s? Going back even further we see mechanical servants in films as early as 1909 in the British film *The Electric Servant*. Or what about the Turk-like humbug in *The Wizard of Oz* from 1939? Hollywood is full of robots, most of which are still beyond our ability to create.

Of course, it didn't begin on the silver screen. Science fiction authors have used the robot as a staple character since the creation of that genre. Isaac Asimov had perhaps the greatest impact on robots in literature with his very human creations and their deeply ingrained Three Laws of Robotics:

1. A robot may not injure a human being or, through inaction, allow a human being to come to harm.
2. A robot must obey orders given it by human beings except where such orders would conflict with the First Law.
3. A robot must protect its own existence as long as such protection does not conflict with the First or Second Law.

Of course, many of his stories involved either creative ways to work *around* these laws or the conflict created by following them.

The robotic creations throughout fiction and movies are a two-edged sword. On the one hand, they inspire people with the promise of technology. I know that I had grand dreams of robots and the future from reading Asimov and others. These works, in whole or in part, inspired me to take a career in technology. On the other hand, these fictions set the bar very high. It is easy to feel let down when you see a robot in the real world if you are comparing it to C3PO or the *Terminator* machines.

Much of the thrill comes back when you build the robot yourself. It may not look like much to others, but you know how much work went into it, the lessons you had to learn and the difficulties you had to overcome to make it work. After all this, when the machine you have crafted works perfectly time and again, it raises itself in your esteem from a few bits of metal and plastic to a status worthy of R2D2. At least until the next project.

FUTURE DREAMS

The big dreams *are* being dreamed in the universities and labs across the world. Japan has been pushing for human-like robotic helpers for years and some companies are showing some excellent results. Honda's P5 and Asimo leap to mind, and Sony's Aibo brought the robot to the home as a pet.

The universities have been hot on the trail of robots and robot minds as well. Rodney Brooks is the director of the MIT artificial intelligence lab, as well as the founder of their humanoid robotics group. The Cog project has been exploring the boundaries of human/robot interaction ever since Brooks dreamed it up in 1993. Newer projects explore other aspects of social robots. Macaco is a dog-like head that Artur Arsenio is using to explore robot vision, while Kismet is used by Cynthia Breazeal's team to explore social and emotional interactions between this infant-like robot and its caretakers.

On a more commercial front, the robot reality of Sarcos in Salt Lake City is getting close to some of the movie dreams. Alvaro Villa's animatronic creations are used in the movies to create more dreams—or in some cases, with his *Crypt Keeper*, nightmares.

And there are more, many more. Scientists are modeling robots on all aspects of nature, basing their work on insects, birds, dinosaurs, and, of course, humans. A visual tour of robots can be found in the beautiful book *Robosapiens*, by Peter Menzel and Faith D'Aluisio. In a way, researchers are still mimicking nature like the automata builders did in the fifteenth century. Only their tools are better, and their creations are more capable.

It makes sense to mimic nature. Time and hardship have sculpted the world's creatures, enabling them to survive and adapt in ways that roboticists still only dream about. Some researchers are not content to simply imitate nature's results, however, but go one step behind the scenes and try to reproduce nature's methods. These researchers are working on evolving robots, using the rules and techniques of genetics to develop machines and the brains to drive them. Machines that were not designed at all, but evolved.

Other researchers are delving down into the very small, not trying to create large complex machines but instead small, simple clouds of machines. Smart dust. Nanotechnology. When we can build our robots on the scale of a single cell, we may have reached the ultimate in robotics. Of course, as stories have told us for hundreds of years, with such power there is also much responsibility and the potential for much harm. This, perhaps, is why Asimov's three laws are so well known and well loved. They remind us that we need to keep our tools safe, so they can be used without causing harm.

Inside Robots

When you open up a robot, what do you see? Mostly the big bits—the outer layers, like the metal or plastic skin, the framework that holds it together, the motors that make it go. Bits of wire. This is what you would see if you could see past your skin into your muscles and bones. So it seems natural to follow this analogy and compare the robot to your body.

As with most analogies, it falls apart if you look too closely. Where do lungs fit into the picture? The endocrine system, or kidneys? What does a robotic liver do? Okay, maybe the kidneys would be the oil filter, and lungs could be cooling fans. But I digress.

A robot is not just one thing, and the study of robotics does not cover just one area of knowledge. A robot brings together systems from many different fields, and to learn robotics is to learn many different technologies.

A robot can be considered in four parts, its frame, mechanics, electronics, and control logic.

The first visible piece is the robot's structure. An animal is held together by its bones, unless you count some of our creepier cousins, in which case the bones are on the outside as an exoskeleton. A robot will have bones of a sort, the parts that give it shape and form. Sometimes the pieces are held together by the robot's shell, or skin. Other times the robot mimics the skeleton model.

The skeleton may give the robot shape but, as in animals, the muscles give it motion. Electric motors make fine muscles, but compressed air and pumped oil are also used to power muscles. For motion, there must also be joints, like your hip or elbow, and a way to attach the muscles to the frame, matching your tendons and ligaments. There are many other pieces to the mechanical puzzle—gears, levers, wheels, and the various forces that they manipulate. All of these fall under this category. Roughly a third of this book is devoted to the technology of mechanics.

With mechanical knowledge in place, we can build automata. To provide sensory input and to send control signals to the muscles, we need electronics. Electronics correspond to the biological nervous system. Your eyes provide visual input, your sense of touch tactile. Internally, we can sense hunger, cold, heat, pain. These signals are routed to our brain along biological wires, our nerves. Commands are sent from the brain to our muscles and organs using these same nerves. Roughly a third of this book is devoted to the electronics that tie the robot together and let it interact with the world around it.

The single greatest thing that separates you from a cockroach (no offense meant) is your brain. Even if your body was that of a giant insect, if you still possessed your mind you would still be you. Remove your head and you become a rather messy meat machine. It is our brain that collects all of the sensory data, organizes it, records it, and then sends out commands in response to it. This is the control center.

Inside of your own head, it probably all feels quite simple and easy. However, this is the hardest of all aspects of robotics and the one that has made the least progress. Artificial intelligence is a new field and it has worked very hard to make its mark in the world. While roughly a third of this book is devoted to the control of the robots, we can't make them intelligent by any stretch of the imagination.

Tools and Supplies

You will get more from this book if you have the following supplies, in addition to your own creativity and imagination.

The mechanical examples are all performed using the LEGO Mindstorms robotic invention system, version 2.0.

Table 1-1   Tools and supplies

Supplies
LEGO Mindstorms Robotic Invention System
Resistors (10 kΩ, 100 kΩ, 10 kΩ potentiometer)
Capacitors (0.1 µF, 1 µF, 10 µF)
Inductors (various)
9-volt battery and battery clip
Solid wire (22 gauge, high-gauge bell wire)
22 gauge solid wire
Breadboard
Solder
Nail, large
Tools
Multimeter
Pliers
Soldering iron
Speaker
Stereo

For the electronics experiments, it will help to have a soldering iron and solder, a prototyping breadboard, 22 gauge wire, small pliers, and a handful of resistors and capacitors, as well as a 9-volt battery and battery clip.

Table 1-1 lists these supplies.

Parting Words of Wisdom

Once you turn to the next page, you will be immersed in a great sea of information. I have tried to make it all as simple as possible and, hopefully, no simpler.

If you find the information too simple, keep going; it gets more complicated later. If you find the information confusing and difficult, remember this. A complex thing is just a collection of simple things working together. Learning how to build a robot is like eating an elephant, you have to ingest it one small piece at a time.

The first few chapters provide a flood of new terms and their definitions. You don't have to memorize them! Feel free to refer back to the chapter as you need.

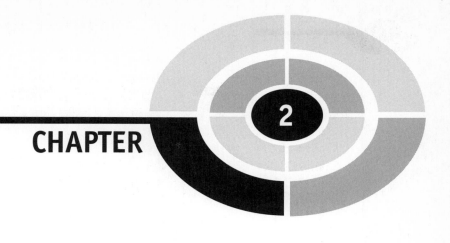

CHAPTER 2

Mechanical Forces

Introduction

Mechanics is not so much the study of the physical pieces of machines as it is the study of the *forces* that machines apply. Force is the result of some physical action and it does *work*.

When you push something, such as a LEGO brick (Fig. 2-1), you are moving it, and this is work. While you are pushing the brick, you are also adding *energy* to it. Once in motion, the brick will remain in motion until some force pushes it in the other direction, taking energy away from the brick until it stops.

If your brick is on a very smooth surface like ice, it continues to move after you stop pushing (Fig. 2-2). This is because the brick still has the energy that you put into it. The motion is in fact energy. A particular form of energy called *kinetic energy*.

But why does the brick slow down and stop? Because it is rubbing against the ice and this takes energy away from it. This rubbing is called *friction*. We talk about all of these ideas in more detail soon.

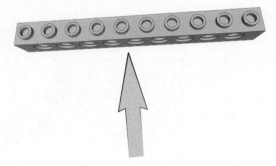

Fig. 2-1. Pushing a brick.

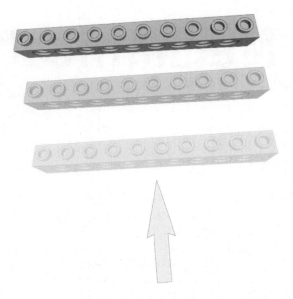

Fig. 2-2. The brick keeps moving on ice.

Even something as simple as a flagpole (Fig. 2-3) is about force. What does the pole do except hold up the flag against gravity, and keep it from blowing away in the wind? Notice how these forces are all exactly balanced and the flag, except for its flapping, doesn't go anywhere.

This chapter introduces the important mechanical forces that we shall be working with. This is an important chapter, because we use these ideas and terms in later chapters as we describe machines.

We also introduce a new way of thinking about numbers. Numbers are just numbers, right? And these numbers are used to describe things like the weight and length of an object, or the time it takes to cook a cake.

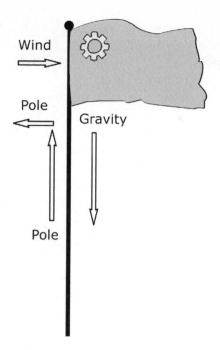

Fig. 2-3. Flagpole forces.

However, when you push something, you are pushing it in a particular direction. On a flat desk, this direction needs two numbers to describe it and together these two numbers, which each measure one piece of the direction, are called a *vector*. Directionless numbers, like time, are called *scalars*.

Energy

There are many different kinds of energy. The two we are interested in here are *kinetic energy* and *potential energy*.

There is no such thing as energy. While an object can *have* energy, nothing can *be* energy. Even light and electricity are not energy.

So what is energy? Kinetic energy is the energy of an object in motion. Kinetic energy is defined in terms of the object's speed, or *velocity*, and how much stuff is in the object, or its *mass*. Mass is typically felt as an object's weight.

Potential energy is when an object doesn't have kinetic energy yet, but could if it fell off the shelf. There are different kinds of potential energy.

This type of energy is stored in the object, as height above ground, in the stretch of a spring, or in a chemical reaction waiting to happen.

We'll come back to these energies in a minute. First we need to define a few more fundamental concepts.

Once we start putting these concepts together into mathematical equations, the names are going to get a bit bulky. Instead of using names, we identify each concept by its *symbol*. We shall be using one- and two-letter symbols for most terms, and sometimes these symbols are even going to be English letters. Greek letters, however, are very popular in math, so hopefully they don't scare you too much. Note that we use the '×' for our multiplication symbol and division is represented by '/' or, more often, by placing the numerator above the denominator like $\frac{n}{d}$.

Not only do we get new symbols to refer to things by, but these symbols are defined in particular *units*, which have their own symbols. You are already familiar with many units. For example, time can be measured in seconds, distance in inches, and weight in pounds.

Units of Measurement

Mechanics is a form of applied physics, and physicists don't always use the units of measurement that we are familiar with. Before we define kinetic energy, let's detour through the basic units we need to know. We look at both the metric and English systems of units here, to get a feel for how they relate. We use metric for all of our actual work.

Units can be confusing. Even the names for the systems of units can be confusing! What we often call *metric* units are more correctly called the *International System of Units*, or *SI* from the French *Système International d'Unités*. *English* units are also known as *Imperial* units, after the British Empire. Ironically, the United Kingdom has now moved almost entirely over to metric. We aren't even going to think about the third system of measurement, *Chinese units* or *Shìzhì*.

POSITION

Everything has a *position* in both space and time. There isn't a unit of measurement for position; however, many of the *other* units relate to the forces needed to change something's position.

The position of an object is based on a measurement, the distance of that object from a fixed point of reference called the origin. Each measurement is taken in a particular direction, called an axis. For example, the distance to the right of the origin might be along the *X*-axis, and the distance behind the origin might be along the *Y*-axis. These measurements make up a coordinate, describing the position of the object. These concepts are explored in more detail when we discuss velocity.

TIME: *t*

Seconds: s

In both Imperial and SI units, time is measured in *seconds*. One second is one tick of your average analog clock, sixty seconds go in a minute, you know. A second.

If you want to get really technical, a second is the duration defined by 9,192,631,770 oscillations of the light from glowing cesium 133 atoms, which can be really tricky to measure.

LENGTH: *l*

(Also, Distance: *d*)
Foot: ft
Meter: m

Closely related to position is *length*, or *distance*. Distance is the measurement in space between two different positions (Fig. 2-4).

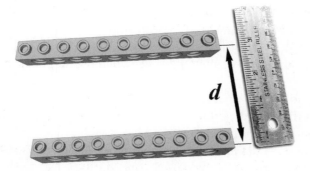

Fig. 2-4.   Distance measurement.

Imperial units define distance as the *foot*, which is in turn twelve inches long. A *yard* is three feet long. Your foot is that thing attached to the tip-end of your leg and, in fact, the original measurement of a foot was based on the average size of feet.

SI units prefer to use the *meter*. The meter was originally defined as 1/10,000,000 of the distance around the Earth. One meter is 3.2808 feet long, or roughly a yard.

Owing to the difficulty of getting an accurate measurement of the size of the Earth by walking around it, the meter was redefined as the distance between two very carefully marked lines on a particular bar of platinum–iridium metal in France.

Today, the meter is defined as the distance that light can travel through a vacuum in 1/299,792,458 of a second. This relates distance *m* to time *s*, since light moves through a vacuum at the same constant speed everywhere in the universe.

One of the great discoveries of science was that light moves away from you at the same apparent speed, no matter how fast you yourself are moving. This phenomenon is described in Einstein's theory of relativity, and isn't something we need to worry about for our relatively slow-moving robots.

MASS: *m*

Pound: lb
Kilogram: kg

The *mass* of something is, roughly, how much stuff it is made of. Technically, the count of the "stuff," or molecules, in an object is the *mole*, used in chemistry.

In the presence of gravity, mass is felt as *weight*. Without gravity, mass can be felt as an object's resistance to pushing. The more mass something has, the harder you have to push to move it. To get a feel for mass, try rolling a bowling ball and then a marble. Neither object has much friction, but each one has a different mass resisting the push.

Imperial measurement defines mass in *pounds*. One *pound avoirdupois*, to get really picky, is sixteen *ounces*. Troy pounds are different. And there are different definitions for ounce, as well. It can be really confusing, but usually when anyone talks about pounds of mass they all mean the same thing, our familiar sixteen-ounce pound avoirdupois.

Mass in SI units is defined by the *gram*. Since grams are annoyingly small for everyday use, we use the *kilogram*, one thousand grams. "Kilo" is the prefix in SI that means "one thousand." One kilogram is 2.2046 pounds.

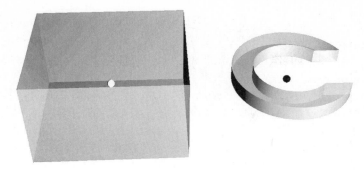

Fig. 2-5.   Center of mass.

The current definition of kilogram is defined by a carefully protected bar of platinum–iridium metal in France. Scientists would like to replace this bar of metal with a more universally accurate definition but haven't been able to come up with a good one yet.

An object has mass spread all over and through it. Since objects can be in all sorts of awkward shapes, it can be difficult to calculate how they will behave when pushed, unless we simplify them.

For every object, there is a single point in space where the mass of the object is balanced in every direction. This is called its *center of mass*. In the presence of gravity, this is also known as the *center of gravity*, or *COG*.

For a sphere, square, box, or other well-behaved and symmetrical object, this point is in the very center of the object. For more unusual shapes, such as the letter "C," it takes a bit more math to find the center of gravity (Fig. 2-5). Once found, however, the object can be treated as a single mathematical point for many calculations.

VELOCITY: *v*

Feet per second: ft/s
Meters per second: m/s

Velocity is a *derived unit*. This means it doesn't represent anything new, but combines *base units* to come up with a new concept. Time, length (or distance), and mass are base units.

Velocity describes the change in position of something over time (Fig. 2-6). To describe velocity you need both a distance, in meters or feet, and the time measurement in seconds. Velocity is also known as *speed*.

A velocity of 3 m/s means the object has traveled three meters in one second. It could also mean it moved six meters in two seconds, nine meters in

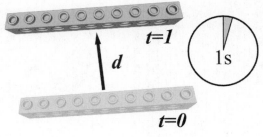

Fig. 2-6.   Velocity.

Fig. 2-7.   Single axis.

six seconds, and so on. You assume the measurement is made across one second unless otherwise specified.

Once an object is moving, it stays moving until another force acts on the object to make it stop moving. This is Newton's first law of motion:

> A body must continue in its state of rest or of uniform motion in a straight line, unless acted on by some external force.

Note that an object doesn't just move, but it moves in a *direction*. If you are pushing beads on a wire, there is only one direction they can move, along the wire. Technically, there are two directions, back and forth along the wire.

This wire, with its constrained direction of motion, can be thought of as *one-dimensional*, since there is just one dimension or direction of motion, along the wire. The wire itself is an *axis* (Fig. 2-7).

When you are pushing blocks on your desk there are two parts to the direction it can move: left/right, also called the *X-axis*, and toward/away from you, the *Y-axis*. This is a two-dimensional space, and the axes are directions along the desktop (Fig. 2-8). Each axis is perpendicular to the other axes.

Distance along a wire is described by a single number, the distance from the start. Distance on a desktop, however, needs two numbers to describe, the distance between two positions along the X-axis and the Y-axis (Fig. 2-9). These are considered part of one measurement, a two-dimensional *vector*. If this distance describes two positions of the same object, it describes the two-dimensional motion $(\Delta x, \Delta y)$.

The Greek Δ, called "delta," represents a change or difference in some value. Therefore Δx represents the change in position along the X-axis. Another way to represent a change in position is to put a single dot over the value, $\dot{x}$.

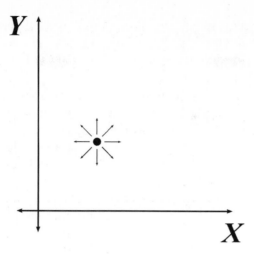

Fig. 2-8.   Two-dimensional space.

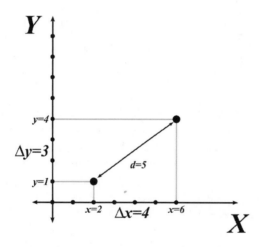

Fig. 2-9.   Two-dimensional motion vector.

Given the distance along X and the distance along Y, you can calculate the distance d between the two points. When the distance represents motion, it is a *vector*. When this motion is considered across a measured span of time, it defines a *velocity*:

$$d = \sqrt{x^2 + y^2}$$

$$v = \sqrt{\Delta x^2 + \Delta y^2} \tag{2-1}$$

If you can lift objects off the desk, there is a third axis that points straight up from the desk, the *Z-axis* (Fig. 2-10), perpendicular to both *X* and *Y*. *Z* adds a third number to the vector, making it a three-dimensional vector.

Another word for each number, *x*, *y*, or *z*, in a vector is *ordinate*. While the motion of an object in space is a vector, the position of an object in space is called a *coordinate* (Fig. 2-11). A coordinate specifies the distance of an object from an arbitrary, predefined point in space called the *origin*. The origin is normally described as being at position $x = 0$, $y = 0$, $z = 0$ or $(0, 0, 0)$ (Fig. 2-10). All positions in space are then distances from this origin.

The whole package, the three axes plus the origin and the units used to measure it, is called a *coordinate system*. In this case, the *Cartesian* coordinate system. There are other systems of position, such as the polar coordinate system, not used here.

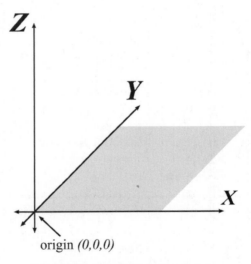

Fig. 2-10.   Three-dimensional space.

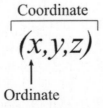

Fig. 2-11.   Ordinates in a coordinate.

ACCELERATION: *a*

Feet per second per second: ft/s²
Meters per second per second: m/s²

Like velocity, *acceleration* is a derived unit. Also like velocity, acceleration occurs in a particular direction. Acceleration is a change in velocity over time. For example, step on the gas of your car or drop a ball and its velocity increases over time. Acceleration can be in terms of meters or feet, since it relies on a distance measurement.

Since acceleration is the change of velocity over time, and velocity is itself a change in position over time, acceleration is the change in position over time over time. This is why we use symbols. It is much easier to say this with the mathematical statement:

$$a = \frac{d}{s^2} \tag{2-2}$$

While velocity is easy to visualize, what exactly is a change in position over time squared? What is square time? This question doesn't have an intuitive answer.

Acceleration isn't motion, but a change in motion. It is the result of a push against an object. Where the position of something might be described symbolically as x, and the change in position, or velocity, is Δx or $\dot{x}$, the change in velocity adds a second dot to become $\Delta(\Delta x)$ or $\ddot{x}$.

Gravity is an acceleration. The force of gravity on Earth is constantly trying to accelerate everything on Earth at a rate of 9.80665 m/s² toward the Earth's center. This acceleration is the gravitational acceleration constant g.

What this means in practice is that if you drop something, it falls. In fact, if you ignore wind resistance by dropping something in a vacuum or by dropping a heavy, round ball, that ball will accelerate at 9.8 meters by each second, squared.

At the start of the fall, it isn't moving and at the end of the first second, it will be moving at the velocity of 9.8 m/s. The average velocity, from zero to 9.8 m/s, is actually half that amount, and the ball has only traveled 4.9 meters.

But it's still speeding up! At the end of the second second, it is traveling at $4 \times 9.8 = 32.9$ m/s. The average velocity is still about half that. Each second it falls, the object gets faster and faster—until it hits ground. Then, typically, it breaks (Fig. 2-12).

Starting from a dead stop, you can calculate how far the ball has fallen using this equation, where d is the distance traveled and t is the time in

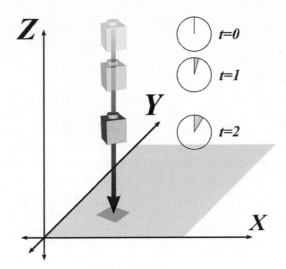

Fig. 2-12. Acceleration.

seconds of the fall:

$$d = \frac{a \times t^2}{2} \tag{2-3}$$

FORCE: *F*

Pounds force: lbf
Newtons: $N = kg \times m/s^2$

Generally speaking, force equals mass times acceleration, $F = m \times a$. Note that this is the "mass" m and not the "meters" m.

Just a few definitions into the list of physical formulas and we already find another meaning for pounds: pounds as a pushing force, the force of whacking a nail with a hammer. The pound is definitely overused and will be ignored here. From this point forward, we will abandon the Imperial units and talk only in terms of SI units.

Let's talk about Newtons, and not the figgy kind. The SI unit of force is named after Isaac Newton. Newton's second and third laws of motion relate to the force applied to an object. Newton's second law is:

Change of motion takes place in the direction of the impressed force, and is proportional to it.

Or, when you push on something, it moves in the direction of the push. How much your push changes the object's motion depends on how hard you push.

Newton's third law of motion is:

Action and reaction are equal, and in contrary directions.

So when you push on an object, such as a bowling ball or marble, you feel an equal force pushing back at you from the object.

A *Newton* is practical acceleration. Pure acceleration is a bit too abstract for solid objects. Sure, gravity can get away with accelerating everything like magic. But when you push on it, a LEGO brick moves differently than a car.

To get the same velocity out of the car, you have to push it a *lot* harder than the plastic brick. The difference lies in the mass of the object. The more mass, the harder you have to push it to accelerate it at the same rate as the lighter object.

As far as acceleration is concerned, it doesn't matter how hard you had to push to get to a speed. But your muscles care. The Newton measures how hard and how long you have to push to get an object moving to a certain speed, or to get it to *stop* moving. It measures *force*.

The other side of force is how much energy a moving ball, for example, transmits to your head when it hits you. (Kids, don't try this at home. No, not even with your kid brother.) A sponge ball traveling at 10 meters per second won't hurt much. A baseball, with more mass than sponge, thrown at the same speed, will hurt. A lot.

It's not just the speed and mass of the object that matters. It's how hard your head had to push against it to accelerate it in the opposite direction and stop it. Not only that, but how *quickly* your head stops the ball makes a big difference.

Being a bit more sensible about catching baseballs, let's use a baseball mitt. If someone throws the ball at you and you hold your hand stiffly so it doesn't move, the ball will stop quickly and it will hurt. If you let your hand travel with the ball a bit, cushioning the blow with the motion, it's much less painful.

In the first case, you reverse-accelerate (*decelerate*) the ball all at once, which requires a lot of force in a small amount of time. In the less painful catch, the same *total* amount of force is needed, but since it is spread out over more time there isn't as much force during each moment. It's like the difference between the pain of pushing the sharp end of a thumb tack into your finger and what you feel from the flat end on the finger doing the pushing. Deceleration is just acceleration, but in the opposite direction of an object's current direction of travel.

MOMENTUM: *p*

$p = \text{kg} \times \text{m/s}$

Momentum is closely related to force. Where force is tied to the acceleration of the object, momentum uses its velocity. Momentum is defined as mass times velocity, $p = m \times v$.

ENERGY: *E*

Joule: $J = \text{kg} \times \text{m}^2/\text{s}^2$

Finally, we have reached the last definition in this section. It's been a thick few pages to this point, but we needed all of the above to finally be able to define *energy*, also known as *work*. The unit of *energy* is the Joule, which is defined as force across a distance:

$$J = N \times m \tag{2-4}$$

Lifting an apple above your head is work and it adds energy into the apple. As the apple is in motion, your hand is accelerating it upward and, as the apple accelerates, it gains velocity and momentum. Once it's above your head and gravity decelerates it to a stop against your palm, it loses its velocity and momentum. However, it has gained *potential energy*. When you drop the apple, all that work you put into it is undone as it regains its velocity and momentum (thanks to gravity's acceleration) until it hits the floor. And, typically, it breaks, making a mess. Be sure to clean up after trying this one.

Kinetic energy: *KE*

$$KE = \frac{m}{2} \times v^2 \tag{2-5}$$

Kinetic energy is the energy stored in a moving object. Note the similarity of kinetic energy to momentum. Kinetic energy is one half of the mass times the square of the velocity. Make an object move twice as fast and it has four times the energy. This is related to the squared time factor from the acceleration that pushed it to this velocity.

Potential energy: *PE*

Where kinetic energy is the energy of a moving object, potential energy is the energy that is hidden in the object's situation, waiting for an opportunity to become kinetic energy.

There are actually different kinds of potential energy. An object held high above the floor and against gravity has a bunch of potential kinetic energy. The amount of potential energy due to gravity is calculated as the mass of the object times the acceleration of gravity and its height above ground:

$$PE = m \times g \times h \tag{2-6}$$

However, there is also potential chemical and electrical energy in a battery, and a different look at potential energy is found in a stretched rubber band or spring.

Let's look at potential energy in more detail, since it provides a way to store energy for later use.

Storing Energy

While an object in motion has kinetic energy, there are ways other than motion to store mechanical energy, such as in *springs*. Of course, you can store types of energy other than mechanical; batteries, for example, use stored chemical energy to move electricity.

There are different types of springs: long flat leaf springs, round disk springs, coil springs that you pull on to stretch and coil springs that you press on, or *compress*, to squish, and even rubber bands. These all do roughly the same thing and store energy by converting the force applied to the spring into potential energy.

The force applied to the spring is called *stress*, and the change it creates in the spring is called *strain*. Force, as described above, is still mass times acceleration.

Strain is the internal change of position of the atoms in the spring. Each atom inside the spring is attracted to the other atoms by the atomic forces. These forces are like gravity, except they are much stronger and are only effective at extremely small distances. Note that there are other atomic forces that push the atoms apart.

Between the pushing and pulling of the atomic forces, it is as if each atom is connected to its neighboring atom by a spring. The spring is made up of millions of smaller atomic "springs," in the same way that an apple is pulled toward the Earth by the force of the "gravity spring." Of course, these aren't tiny physical springs inside our big spring. They are the forces that keep our universe together.

Strain occurs when you pull or push the atoms in an object against the atomic forces that are working to keep them in place. The object then changes shape, or *deforms*. This deformation is the visible manifestation of the strain inside an object. You feel the strain by the force of the object trying to return to its original shape.

Of course, if you put too much strain into something, the atoms can be pushed and pulled too far and break their internal springs. Then the object bends or breaks, and its stored potential energy is rearranged so we can't use it anymore.

When you stretch or compress a spring, it pushes back with the force of:

$$F = -k \times l \qquad\qquad (2\text{-}7)$$

k is the *spring constant*, which describes how "springy" the spring is, and l is how far the spring has been stretched or squished (Fig. 2-13).

The spring's force pushes in the opposite direction of the force used to deform it. The spring's stored force exactly matches the force it took to deform it. The same with a rubber band, though rubber bands are only useful in the stretch direction.

In robotics, you can use the force stored in a spring or rubber band to push against the acceleration of gravity, to make it easier for the robot's motors to lift something. Springs can also be used to hold something in position, to keep it from moving.

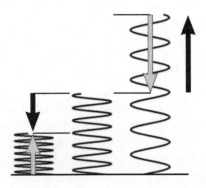

Fig. 2-13.   Spring forces.

Losing Energy

Energy is never really "lost," it is only shuffled around into different forms. Kinetic energy is converted into potential energy, which might be stored as deformation in a spring or as a displacement against gravity. These might eventually be turned back into kinetic energy. So energy changes form, but it never goes away. So why does the brick stop moving when you push it across your desk? Where is the kinetic energy going?

The energy you are adding to the brick is being lost into the desk through *friction*. Friction is what we call the process of transferring energy by rubbing things together. The force of friction always appears to act in the opposite direction of the kinetic energy causing the friction. Because of this, an object rubbing against another object slows down and stops.

There are two kinds of friction. An object at rest on a surface has a higher friction than an object that is already moving. The stationary, or static, friction force is called *stiction*. When you press on the brakes in your car, the force holding the wheel to the road is this stiction. Once you start skidding, the lesser force of friction takes over. Once the wheels start to get hot and melt, it gets even more difficult.

Friction can be an extremely complicated thing, especially if either of the objects being rubbed together are sticky, fuzzy, oily, or otherwise not dry and smooth. Plain vanilla friction is molecules, by way of their atomic forces, banging against each other. Let's look down into the atoms of the object and see what this banging means to them.

Everything that you come into contact with has *heat*. Heat is a form of energy and, in fact, it is a form of energy stored by motion. Each atom within an object is not just sitting there quietly but is vibrating madly against the springy forces that keep it from escaping. This vibration is described as the heat of the object. The faster the atoms in an object are moving, the more energy it is storing; it's hotter.

Of course, if something gets too hot, the atoms get enough energy to break their springs and they escape. At one temperature the springs are only mostly broken, so the object doesn't hold together as firmly and its atoms are pulled down by gravity. It *melts*.

Get it even hotter and the atoms have enough energy to jump up away from gravity's pull and the attraction of its neighbors and the object *evaporates*, or becomes a gas. Of course, like almost everything else in this book, it's a bit more complicated than that. But you get the general idea.

When you hold a hot object against a cold object, the atoms in the hot object bang into the atoms in the cold object, making the slower atoms move faster and the faster atoms move slower. Some of the energy is transferred so that both objects come closer to having the same amount of energy. It is as if the heat flows from the hot object to the cold one.

When you push a block across the table, the atoms in the block are banging against the atoms in the table. This takes some of the kinetic energy, which is an energy of motion on a large scale, and transfers it into the motion of the individual atoms in the table.

These microscopic interactions also make the atoms in the block vibrate faster in their springy cages, so some of the block's kinetic energy is transferred to its own atoms as heat. Both the desk and the block heat up and kinetic energy is lost. The total energy is the same, it has just changed from one form (motion) to another (heat).

Even when you squeeze or stretch a spring there is friction, only this friction is inside the spring. Some of the force you apply to the spring turns into heat as the spring moves. You can feel this internal friction yourself with a coat hanger. Get a wire coat hanger that nobody will mind you breaking. Now, grab it firmly in both hands and make a sharp bend in it. Now bend it back the other way. Do this a few times quickly, and the bend will become very hot. Do it enough, and the hanger breaks.

We normally want to preserve the kinetic energy of an object and keep it from being converted to heat. Motion is useful and, unless you are cooking something, heat is not. There are different ways of doing this, such as lubrication or ball-bearings. We'll look at these friction-reducing ideas later.

Summary

In this chapter, we introduced a bunch of mechanical forces, their meaning, and symbols to represent them. We learned how to describe something's position in space and its distance from another object. We learned about the energy in moving objects as well as the force and energy needed to change an object's motion, not to mention the difference between force and energy. Finally, we took a look at storing energy in springs and how energy is lost through friction.

Quiz

1. If a ball dropped on planet Earth takes a half second to hit the ground, how high was it dropped from?

Bonus: If you drop a ball from 3 meters (about 9 feet) above ground, how long does it take to hit?

What does it look like if you graph the results of these tests? One graph would show how far the ball falls for a range of times. Another graph would show how long the ball takes to fall a given distance.

2. Take a half-kilogram ball (about a pound) and throw it at 10 meters per second at a wall (NOT someone's head). Assume the wall takes 1/10 of a second to stop the ball. What is the force applied to the ball over that tenth of a second? What if the wall was really soft and took 1/4 second to stop the ball?

3. What is the kinetic energy of this half-kilogram ball moving at 10 meters per second? What if it were moving at 20 meters per second? Try graphing the kinetic energy of the ball at different velocities.

4. If you carry this poor, abused half-kilo ball to the roof of your building, which is 4 meters high, how much potential energy does it now have?

What does all this have to do with robots? Lots! Say you had a 10 kg robot that was trying to climb a 10 degree hill. How much push (acceleration) does it have to have to make it to the top? If the hill is 100 meters high, how much energy was expended to get it there (think potential energy)? You don't have enough information to answer these robot questions yet, but you will soon.

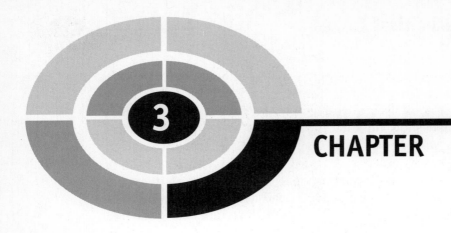

CHAPTER

3

Simple Machines

Introduction

Now that you have had a chance to meet the relevant physical forces, we can start playing with them. The basic set of mechanisms for manipulating force are known as the *simple machines*. Variations on these simple machines are the base components used by more complicated machines. Robots, for example, are built up from many of these simple machines working together.

Before looking at moving mechanisms, we look at the strength of structures that don't move. There are two branches to mechanics. One is *statics*, which is the study of objects that aren't moving. Though there are forces applied to a static object, those forces are balanced so there are no continuing changes in shape or position.

The other branch is *dynamics*, which is the study of objects in motion. Dynamics is further divided into *kinematics* and *kinetics*. Kinematics is the study of mechanical motion without worrying about the forces behind it. The shape of the motion. Kinetics then studies the forces driving the motion.

In this book, we are not going to be concerned with this breakdown, though most of our work is related to dynamics.

After a brief introduction on how to build models that hold together instead of fall apart, we investigate some of the variations available in ramps, or *inclined planes*, levers, and wheels. These devices, and the others that are related to them, all provide methods of trading speed for strength.

For example, how is it that one person can lift a car engine with their hands? On the one hand, we have an engine that weighs, for example, 200 kg (over 400 lb); on the other hand, an average person who is going to be able to lift a quarter of that amount, at best.

The person could lift that engine, though, using simple hand-powered machines. Remember that force operates over time, and energy is the application of force across space. What we do is use a small amount of force over a long period of time and a long distance. We can add up this long, drawn-out force up and use it to lift an engine a short distance.

Essentially, what the simple machines do is let us trade distance for force. This is also known as using *mechanical advantage*. As a side effect, we trade time for force. Moving something twice as far to get twice the force at half the distance is also going to take more time.

So, read on to learn how to convert distance and time into superhuman strength. We also introduce a few more terms as we need them.

Structural Strength

There is a whole study of the strength of materials and structures known as mechanical statics. Engineering classes spend a lot of time on this subject, since we prefer our bridges to remain standing and our buildings vertical. For our purposes, however, there is one simple principle to learn: *Triangles are strong*.

Everything else flows from this, even if it's not obvious. Let's look at triangles and squares in general. Then we can look for the hidden triangles in a few structures.

TRIANGLES AND SQUARES

Different shapes have different strengths, just because of the shape. A shape will also have different strengths depending on how force is applied to it.

For example, look at your basic drinking straw. If you squeeze it from end to end, pushing through the straw's length, it's pretty strong. But grab both ends and bend it and it folds up with no effort.

For another example, look at a ruler. If you place it flat on your desk so that half sticks out over the edge, you will find that you can easily bend the ruler down. However, turn it so that the skinny edge is against the desk and you will find it much harder to bend down. In fact, it will twist before it bends.

To look at the intrinsic strength of squares and triangles, we need to build each of these shapes with hinged corners (Fig. 3-1), so that the corners can bend.

If you push down on a square, straight down on the top, it's pretty strong. The force is sent down the sides and into the ground, or bottom edge. The force passing straight through the sides is called *compression* and the rods are very strong under compression.

If the force goes to one side even a little bit, the corners bend and the shape folds up flat. It has no built-in strength in the side-to-side direction.

When you push on the point of a triangle the force is passed down the center of its edges, compressing them, and into the joints at the other side. At the other side the force pushes out at the joints, which pulls on the bottom bar. This pulling force is *tension*, and these bars are also very strong under tension.

At no time do the joints of the triangle rotate and let the shape collapse. The edge rods or pivot pins will break first. This is what I mean when I say that triangles are strong. The shape has strength built into it.

Another strong shape is the circle, because force at one edge tends to be redirected around the edge into compression. You see circles in action in arches, eggs, and domes.

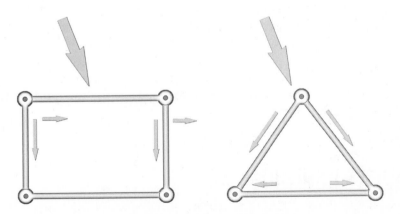

Fig. 3-1.   Strength of shapes.

A third force, not illustrated here, is applied by scissors to paper. This is the *shear* force. I mention it only to be complete. The forces, compression, tension, and shearing, all cause stress in an object.

There is a lot that can be said about the static strength of structures. We looked at just two cases using a hinged link. There are other shapes, and other ways of fastening the pieces together. And then there is the math behind it all. However, we can safely ignore most of those details.

HIDDEN TRIANGLES

When you nail two long boards together with one nail, that nail makes a nice pivot point, allowing the two boards to rotate at that point, resisted only by friction. But what happens when you put two nails into the boards? This keeps them from rotating, so these nails should act to make the square strong. Doesn't this break the rule of triangles?

No, because that second nail defines a triangle (Fig. 3-2). When you get to the section on levers, you will see that the strength provided by that nail triangle is actually pretty small.

Your house is probably made up of squares, and yet it doesn't fall down on you. What makes a wall strong? The wall's covering, sheetrock on the inside and plywood and decorative materials on the outside. This covering

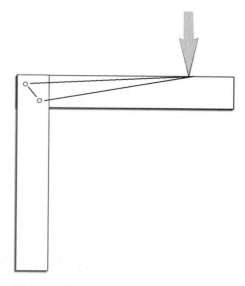

Fig. 3-2.   Hidden triangle in two boards.

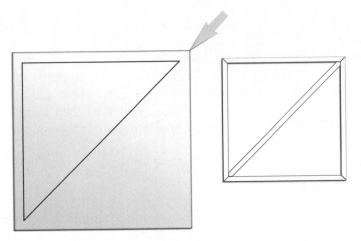

Fig. 3-3.   Hidden triangle in a wall.

defines giant triangles in the wall. An open wall with just a diagonal brace will be almost as strong as a covered wall (Fig. 3-3).

There are times when you won't find a hidden triangle in your structure, such as in the construction in Fig. 3-4. This is when you look for levers.

When you build your machines, you want to do so with a keen eye to the triangles and levers in it. Engineering statics is a complicated subject. Simplifying it down to triangles and levers, however, makes it about as simple as it can be while still being useful.

Inclined Plane

The *inclined plane*, a simple ramp, is the simplest of the simple machines. You use it when you walk up a hill. If you have a large hill in your town, think about how much easier it is to walk up a path to the top than to have to climb straight up a wall of the same height. This is the inclined plane in action.

There are different ways to think about the inclined plane. We shall look at how it helps us raise objects against the acceleration of gravity first.

Starting with an object on the ground, gravity is accelerating it down at the brisk rate of $9.8 \, \text{m/s}^2$. The force developed by this acceleration is directed straight down into the ground (Fig. 3-5).

What happens when you put that same object on a slippery slope (Fig. 3-6)? Assuming there is no friction, it will slide down to the bottom of the ramp. While gravity is accelerating the block straight down, the ramp is in the way.

Fig. 3-4. Hidden lever.

Fig. 3-5. Gravity pushing a block down.

At this point, the force from gravity is split into two parts. Each of these two parts describes a direction.

The force of gravity is directed down, which makes it the force direction. This acceleration can be described by its own vector. In our simplified, two-dimensional world, the X ordinate in the vector is the amount of acceleration

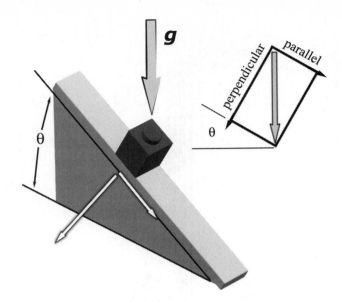

Fig. 3-6. Block on a ramp.

side to side. The *Y* ordinate is the acceleration straight down. The overall length of the vector gives the strength of the force. In this case, the gravity vector would have a length of 9.8, representing gravity's $9.8\,\text{m/s}^2$ acceleration.

Now for the two force vectors as applied to the block on the ramp. One force vector, the *perpendicular* vector, points directly into the ramp's face and creates friction. The other force vector, the *parallel* vector, points along the face of the ramp and is the force that moves the block down to the ground.

The relative size of each force vector depends on the angle of the ramp, θ (theta, sometimes shown in the capitalized form Φ, and which often represents an angle). The perpendicular force vector has a length of $g \times \cos\theta$ and the parallel vector a length of $g \times \sin\theta$. The sine function starts at zero and increases to one as the angle increases to 90 degrees. When the object is flat on the ground, the parallel force is zero. If the ramp is in fact a wall (Fig. 3-7), the parallel force is one. The cosine function is the reverse, starting at one.

For a ramp at a 30 degree angle, the perpendicular acceleration is $8.43\,\text{m/s}^2$ and the parallel acceleration is $4.9\,\text{m/s}^2$. Most of the force is being applied to friction, but still half of it is applied to moving the block.

Wait a second. Those numbers don't add up to $9.8\,\text{m/s}^2$! Right, because the parallel and perpendicular vectors don't add up to the same length as the gravity vector. Instead, they are appended to each other, making half of a box around the gravity vector. If you fill in the box, as shown in the corner of Fig. 3-6, you can see the relationship. Gravity cuts the corner across the box.

Fig. 3-7.   Block on a wall.

The relationship is described by the Pythagorean theorem, which states:

$$a^2 + b^2 = c^2 \qquad (3\text{-}1)$$

where a and b are the lengths of the parallel and perpendicular sides, and c is the hypotenuse, or long side described by gravity in our example. Other ways to show this relationship are:

$$c = \sqrt{a^2 + b^2} \qquad (3\text{-}2)$$

$$a = \sqrt{c^2 - b^2} \qquad (3\text{-}3)$$

Note that this is the same relationship used to calculate the distance between two points, used in the section about velocity in Chapter 2, and equation (3-2) is the same as equation (2-1).

Now, back to the ramp and how it changes our use of force. When you push an object across the floor, you only have to counteract its friction and momentum. Ignoring friction, we are only acting against the mass of the object. Lifting the object straight up, we act against the full force of gravity. When we push the object in the direction of the ramp, we are only pushing against the parallel force. Remember that we are ignoring friction, which is a function of the perpendicular force.

For a very shallow ramp angle, there is almost no parallel force, and it is easy to push the object up the ramp against gravity. However, it doesn't get very far off the ground unless we push it a long ways. Lifting straight up, all of the motion is up but it's much harder work. This is where we are trading side-to-side motion for force. The farther we push an object to raise it a certain distance, the less force we have to use pushing it but the longer we have to push.

WEDGE

A *wedge* is essentially an inclined plane in portable form (Fig. 3-8). Instead of gravity pushing, we apply our own force.

Looking at the wedge, if its sides are at 45 degrees to the towers, then for every meter we push it into the gap, it pushes outward by one meter. This just redirects the force and doesn't make the pushing any easier.

If the wedge is long and skinny, however, so that for every two meters you push it down it pushes out one meter, it has a *mechanical advantage* of 2.

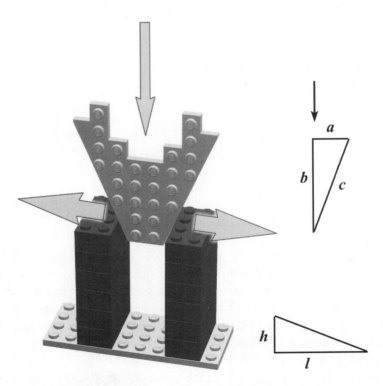

Fig. 3-8.   A wedge for mechanical advantage.

You have to push it twice as far, but it generates twice as much force outwards.

In the diagram in the corner of Fig. 3-8, you can see that the wedge is just a ramp on its side. The height of the ramp h is the width of the wedge and the length l is the length of the wedge. The mechanical advantage (MA) is defined as:

$$MA = \frac{l}{h} \qquad (3\text{-}4)$$

In terms of energy, we haven't done anything except rearrange some terms. Remember, energy is just force in Newtons applied across distance. We have the energy we are putting into the wedge, E_{in}, and the energy the wedge is putting into the towers, E_{out}. The relationship is defined as:

$$(MA \times E_{out}) \times m = E_{in} \times (MA \times m) \qquad (3\text{-}5)$$

where m is distance. On one side, the mechanical advantage adds to the output force. On the other side, we have to move the wedge that much farther.

Another way to look at it is as:

$$E = (m \times MA) \times \left(\frac{F}{MA}\right) \qquad (3\text{-}6)$$

where E is the energy, or work being done, and m is still distance. The mechanical advantage MA just shifts the focus of our efforts from the force being applied to the distance we have to move to apply it.

The wedge is a different way of looking at the inclined plane. A third look is the screw.

SCREW

A *screw* is an inclined plane wrapped around a tube. We see the screw shape in bolts, screws of course, and worm gears.

The screw puts the inclined plane into a very compact form. The angle of the plane can be low and we can get a lot of force out of it. This force is applied by turning the screw, which is essentially the same thing as moving a block up the ramp or driving a wedge into a gap.

Einstein said that all motion is relative, and whether the resistance (the block or tower) moves, or the inclined plane (ramp, wedge, screw) moves, the effect is the same.

Levers

A *lever* is another way to apply mechanical advantage. Though the lever is a more complicated machine, it can be described with simpler math than the inclined plane.

LEVER MACHINE

 Try This: For levers, let's build a LEGO lever machine. In the fine tradition of these types of projects, the construction steps are depicted in pictures, shown in Fig. 3-9. The last step, and an informative example showing range of motion, is in Fig. 3-10. Pushing down on the long arm of the lever creates an upward force on the short arm. While the long arm moves four times as far, the short arm creates four times the force.

You can hang weights at different points on this machine to get a feel for how it works. How does it feel when you hang the weight on the long end

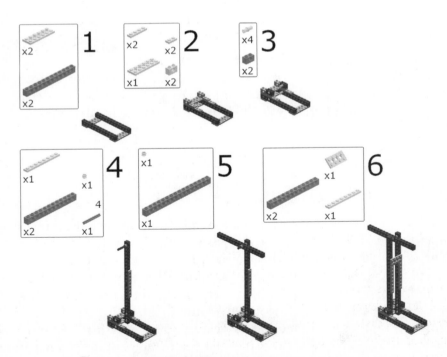

Fig. 3-9. Building the lever.

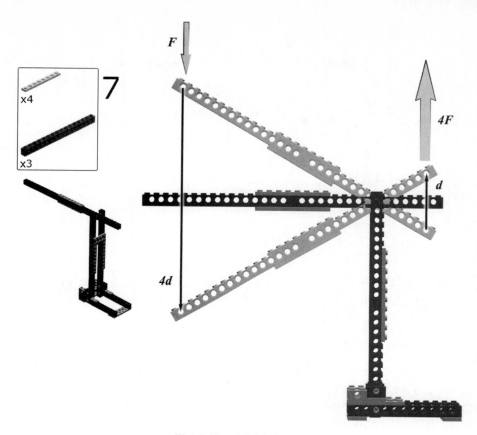

Fig. 3-10. LEGO lever.

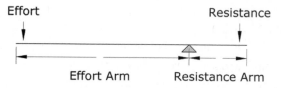

Fig. 3-11. First-class lever.

instead of the short end? Try the weight at different positions and see how it affects the force and range of motion.

A formal description of the lever is given in Fig. 3-11. This lever is known as a *first-class lever*.

All levers have the same three pieces. The pivot in the middle is the *fulcrum*. The distance from the fulcrum to where the input force, or *effort* (E), is applied is the *effort arm* (d_E). The distance from the fulcrum to where the force is used is the *resistance arm* (d_R). The lever is applied against *resistance*

Fig. 3-12. Second-class lever.

Fig. 3-13. Third-class lever.

(R). If d_E is larger than d_R, then your force is increased. The mechanical advantage is calculated as:

$$MA = \frac{d_E}{d_R} \tag{3-7}$$

Equation (3-7) is essentially the same as equation (3-4) for the wedge.

Though we look at several different simple machines, each with their own approach to increasing force, the principles behind each one are the same. Force is force, and the rules governing force are not going to change. If we seem to be getting more force at the output of a mechanism, somewhere inside it we are spending velocity or distance to get it.

A *second-class lever* is shown in Fig. 3-12, and a *third-class lever* in Fig. 3-13. These have the same pieces as a first-class lever, just in different positions. Note that the third-class lever has you pushing harder to get greater distance, backward from the first and second-class levers. Your arm is a third-class lever with your elbow at one end, your wrist at the other, and the muscle attached in between.

Pulleys

A *pulley* is essentially a wheel. Where a wheel wears a skirt of rubber so it has a lot of friction, the pulley has a groove for a rope around its edge. Pulleys are designed so they don't add much friction to the machine they are a part of. Methods for removing friction are explored in a later chapter.

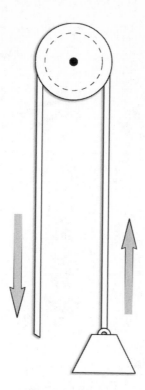

Fig. 3-14. Pulley.

A pulley provides an easy way to change the direction of motion, as shown in Fig. 3-14. When you pull down on the rope, the force is used to lift an object. The pulley itself is fixed into position so that it can rotate but not travel.

PULLEY MACHINE

Try This: Building a pulley machine starts out the same as for the lever machine, as shown in Fig. 3-15. Instead of a pivot for the lever, however, we put in an anchor to tie our string to. We don't use this anchor yet. The pulley itself snaps onto the other side of the tower, as shown in Fig. 3-16.

You can experiment with this pulley, though its operation is fairly obvious. The fun begins when we begin to make pulleys behave like levers, to amplify our strength. To double our strength, we need to find a way to pull twice as much rope while making the weight rise the same distance. This could look like Fig. 3-17. The new pulley is not fixed into position, but travels with the

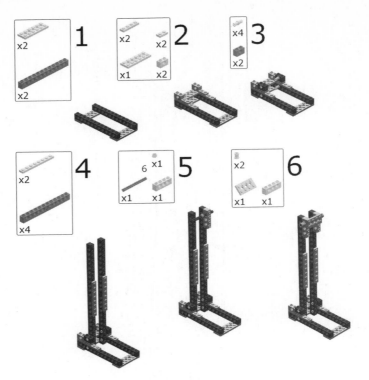

Fig. 3-15. LEGO pulleys.

Fig. 3-16. Single pulley.

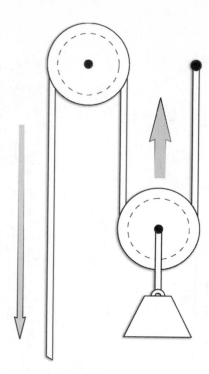

Fig. 3-17.   Double pulley.

weight. The weight helps to keep this pulley on its rope. The end of the rope is anchored somewhere near the fixed pulley.

When you pull down on the rope you have a two-to-one mechanical advantage. The mechanical advantage is the number of times the rope goes between pulleys, or the pulleys and the anchor point (not counting the rope you are pulling). In this case, we have two passes. Your pull on the rope must be divided between these two sections, doubling your force and halving the weight's travel distance per pull. You can build a weight for the LEGO pulley as shown in Fig. 3-18. The wheels on the little cart provide some heft. To get an even better feel for the action you can replace them with metal washers or some other heavy thing.

Tie a length of string to the anchor bar (it may be easier to do this if you remove it first). Run it straight down and under the lower pulley. Run the string back up and over the top pulley. When you pull the string down, the weight will rise up.

You can add another pass to the system, giving you a mechanical advantage of three (Fig. 3-19). To test this, you need to add another pulley to your machine.

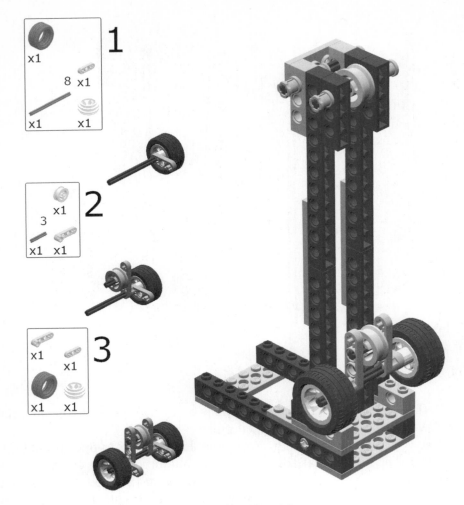

Fig. 3-18. LEGO weight.

Note that our load cart has the two pulleys on the same shaft. Putting pulleys together like this makes a *pulley block*, and it's an easy way to stack pulleys to get greater mechanical advantage. If you added more loops, you would add pulleys to the top of the machine, too. With each loop you reverse the direction that you need to pull to raise the weight. It's normal to have an even number of loops and pulleys so you pull down to raise the weight up.

This combination of pulley blocks and rope is known as a *block and tackle*, where the block is the set of pulleys and the tackle is the rope.

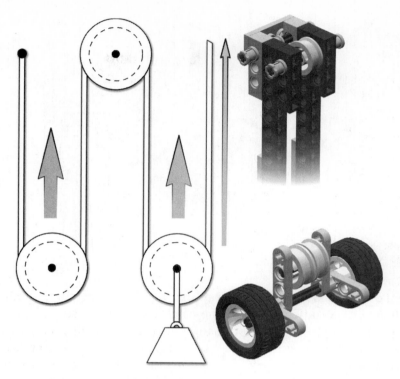

Fig. 3-19.   Triple pulley.

Wheels and Torque

$$\tau = \text{kg} \times \text{m}^2/\text{s}^2$$

We already mentioned that a pulley is a form of *wheel*. Left to itself, a wheel's job is to roll. The wheel is a great invention that we use to reduce the friction of moving objects. It's much easier to roll a car on its wheels than to drag it on its frame.

Not all wheels simply roll. Some wheels push. And, of course, the ground pushes against them. What do these forces look like (Fig. 3-20)? Where the rubber meets the road, a powered wheel pushes against the ground. This force moves the wheel, and whatever it is attached to, forward. Or, in some cases, the wheel could be fixed and the "ground" could be forced to move—it's all relative. The distance from the center of the wheel to the edge is called its *radius*, represented by the symbol *r*.

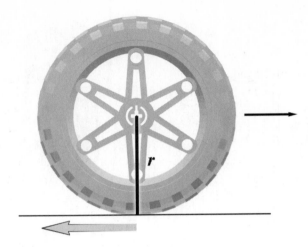

Fig. 3-20. Wheel.

If you draw a line from the center of the wheel to the ground, it looks a lot like a first- or second-class lever. The catch is that the point of effort on a powered wheel is at the same spot as the fulcrum, or pivot point. This gives an effort arm of zero, and yet the wheel moves. If we try to spin the wheel from its edge, the point of resistance is at the center. A zero resistance arm is even harder to solve for in our mechanical advantage equations.

So how do we deal with this impossibility? We add a new force. The force of a wheel turning around a single point is called *torque*, and is usually represented by the Greek letter τ (tau).

The raw calculation of torque was given at the head of this section, $\tau = \text{kg} \times \text{m}^2/\text{s}^2$, which is the same relationship that defines energy in Joules. In other terms, it is $\tau = \text{m} \times (\text{kg} \times \text{m}^2/\text{s}^2)$, also known as $\tau = m \times F$. The m is for the radius of the wheel in meters, and F is the force applied at the edge of the wheel.

The same equations work if the wheel isn't round but is in fact a lever attached to a shaft. The wheel is a continuous, rotating lever. If you push on the edge of a wheel with a force F, the torque you create depends on the radius of the wheel:

$$\tau = F \times m \qquad (3\text{-}8)$$

If the wheel is being powered with a given input torque τ, the push at the edge of the wheel is:

$$F = \frac{\tau}{m} \qquad (3\text{-}9)$$

where m is still the radius in meters.

Gears and Sprockets

A *gear* is a wheel with teeth in it so it won't slip as it rubs against another gear. Two or more gears connected together make a *gear train*. Gears may also push against a toothed bar, called a *rack*, making a *rack and pinion*. The gear is called a *pinion* in this application (Fig. 3-21).

Two gears in a train convert torque from one shaft to force where the gears meet, and back to torque in the other shaft. This makes the gear train a rotary lever, with its mechanical advantage calculated from the radius of the gears. Any number of gears can be paired together like this, in many clever arrangements. Some of these are explored in Chapter 9.

When you select gears and calculate the mechanical advantage of a gear train, you don't use the gear radius, you use the tooth count. The number of teeth on a gear is directly related to its radius by way of the circumference. The circumference of a circle, including gears, is:

$$c = 2 \times \pi \times r \tag{3-10}$$

The Greek symbol π (pi) is a "magic number" that represents the ratio of a circle's circumference to its diameter. Pi has a value of about 3.14, though the numbers after the decimal point never come to an end. The *diameter* of a circle is simply the distance all the way across, or twice the radius.

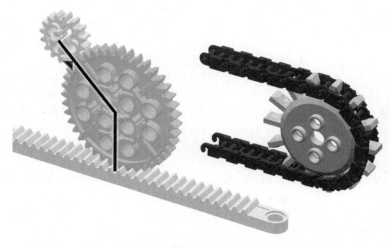

Fig. 3-21.   Gears and sprocket.

The mechanical advantage generated by two gears is the ratio of the number of teeth on the output gear A to the teeth on the input gear B:

$$MA = A : B$$

$$MA = \frac{A}{B} \tag{3-11}$$

Since the number of teeth is directly related to the gear's circumference,

$$MA = \frac{C_{\text{out}}}{C_{\text{in}}} \tag{3-12}$$

If you stretch the circumference of the gear out flat, you have a rack. While it is flat, it is easy to see that the number of teeth you can fit onto the circumference depends on the width of the teeth and the distance between the teeth.

The size of the teeth is called the *pitch* of the gear. The larger the teeth, the stronger they will be but the less you can fit onto the gear. The pitch on two meshed gears must be the same. The shape of a gear's teeth is specially designed to give the gear a smoothly adjusting contact between the meshed teeth at all times.

Many gears can be squeezed into a small space, often with two gears sharing a single shaft. These gears can change huge amounts of distance, in terms of the rotation of the input gear, into huge amounts of torque on the output gear.

A *sprocket* is a gear designed to mesh with the *links* of a *chain* instead of with another gear. The chain travels through space and meshes with a different sprocket. Sprockets and chains are one method of sending force a long way away. They are commonly used on bicycles.

Sometimes a pair of pulleys are used like sprockets, with a tightly-fitting rubber belt stretched between them instead of a chain. These are easy to make and are used on power tools and other equipment to transmit force. Pulleys and belts are quieter than chains, though they can slip. Sometimes you want things to be able to slip, so if the machine gets stuck force is lost in the slipping instead of breaking your machine.

Inside your car you can see a pulley and belt arrangement. The pulley and belt both have large, square teeth. These prevent the belt from slipping, so you get many of the best features of chains and toothless belts. These toothed belts are often called *timing belts*, while the toothless ones are *v-belts*, since they tend to have sloping edges giving them a "V" shape.

Summary

In this chapter we looked at many ways to make yourself stronger. The simple machines let you turn a long, easy push or pull into a short, yet powerful, force that can lift or move an object.

All of the simple machines are based on this principle of mechanical advantage. The simplest machine is just a hill, or inclined plane. A portable inclined plane is a wedge. A spinning wedge is a screw.

A different approach to mechanical advantage is given by the lever and by pulleys. Wheels, we learned, are like rotating levers and have their own force, torque, assigned to them.

Gears and sprockets are toothed wheels that can be combined in large numbers to give a lot of mechanical advantage in a small space.

Quiz

1. If you wanted to study the forces acting on your robot when it is standing still, what would you study? How about when it is in motion?
2. What force is involved when you cut a sheet of paper with scissors? When you crush a grape? When you break a string?
3. Remember that robot trying to climb a 10 degree hill? As it is being pushed up the hill, how hard is gravity pulling against the push?
4. What kind of lever is your knee a fulcrum of? Your ankle?
5. You have a gear with twenty-five teeth, each of which takes 1/8 of an inch (eight teeth per inch). How big is your gear? If you mesh it with a gear that has a 4.97 inch diameter, what is the mechanical advantage?

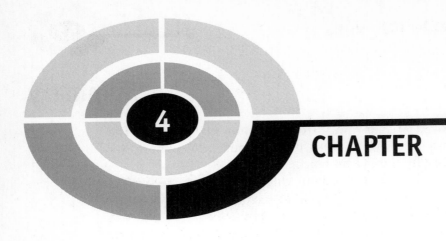

CHAPTER

4

Electricity

Introduction

The study of *electricity* and *electronics* is the study of the electron and proton, the fields surrounding them, and how we affect their behavior. The electron and proton are both *charged particles*, since they carry an electric charge.

The electron is the key player in most electronics studies, since in metals it is the electron that moves around. A charged particle at rest has an electrical field around it, radiating into space. A charged particle in motion generates a magnetic field. These fields are normally studied together as the electromagnetic field, since they are parts of the same thing.

Electronics are the nervous system of the robot, replacing the mechanical gears and cams of the automata with wire and whizzing electrons. This chapter introduces you to the electrical and magnetic forces that are used in electronics. We'll try to keep it short and painless.

Pieces of Matter

Everything is composed of atoms, which are little specks of matter. Atoms are themselves composed of a heavy core of *neutrons* and *protons* surrounded by a whirling cloud of lightweight *electrons*. And that is all a lie.

In the early days, mankind looked around and saw that the world was made up of hundreds, thousands!, of things. Rock and sand and water and mildew and hair and skin and bugs and teeth and metal and wood and bark and leaves and fire—so many things! And of course, we gave everything names and thought up stories as to why these things were here and what they were for.

At some point, philosophers and magicians and alchemists got to thinking about these things. What is wood made out of? What if we took a piece of wood and chopped it into smaller and smaller pieces until we had the smallest possible piece of wood? Is that piece still wood? What does it mean to be "wood"?

At a later point in time, European alchemists would have said that the smallest pieces of everything were composed of the four elements Earth, Air, Fire, and Water. From China, the answer was more likely to be the five elements Water, Fire, Metal, Earth, and Wood.

A more detailed response says that the fundamental piece is the *atom*, which is simply Greek for "can't be cut," and that there are a bunch of different types of atoms. These atoms can be assembled into *molecules*, which in turn are the building blocks of our daily "stuff." Chemistry comes to the front line now and defines the behavior of the various atoms and molecules.

We ultimately identified 116 atomic elements and organized them in the periodic table of elements according to their weights and behaviors. Briefly, in 1999, Berkeley Lab scientists thought they had found element number 118, but later retracted this claim. Elements 113, 115, and 117 are implied by the table but at this time remain undiscovered. At the atomic level, all matter is built up out of these elements and their variations.

Life was good. We had atoms and we had mysterious "forces" like gravity and electromagnetism to keep things in place. Looking deeper, we were able to pry the atom apart into three pieces. Almost all of the atom's mass came from the heavy particles called *neutrons* and *protons*. Neutrons are neutral, in that they do not have any electromagnetic charge. Protons, however, have a positive charge. The neutrons and protons are bound together in the center, or *nucleus*, of the atom.

Whizzing around this nucleus is an array of lightweight *electrons*. One electron has a negative charge equal to one proton, but it takes about

1,800 electrons to make up the mass of a single proton. Early models of electrons showed them in orbits like planets, though later models assign the electrons to mathematical clouds of probability around the nucleus.

The negative charge of the electron is strongly attracted to the positive charge of the proton. Electrons, however, repel other electrons and protons repel other protons. The protons are kept in the nucleus by even stronger forces, but these stronger forces don't leak out so we can ignore them from here on.

There were still questions. There always are. What keeps the electron from collapsing into the nucleus? How do the charges really work? As we dig deeper, our simple atomic model turns into the complex quantum model with its dozens of flavors of quarks and leptons with their colors and flavors and favorite movies.

And the harder we dig trying to make sense of gravity and electromagnetism, the farther the answer slips away. Now we are looking at eleven-dimensional universes composed of ethereal strings. Or maybe it's something else now. We keep looking.

I suspect that, once all is said and done, it will become simple again. But for now, if you want to understand how matter works on a basic level, you must be prepared to have your mind bent.

We, however, are just trying to build robots. So we work with the lie, the convenient simplification. The atomic model of electrons, protons, and neutrons as held together by gravity and electromagnetism.

Electrons in Metal

All metals behave more or less the same when it comes to electricity, so let's look at copper as an example.

A single copper atom has 29 electrons in four orbital shells. The outside shell carries one lonely electron. When you stick a bunch of copper atoms together, you get a useful metallic copper. This can be pulled into wires, rolled into sheets, cast into shapes, and so forth.

That lone electron in the outer shell isn't tied particularly tightly to the copper atom. When the copper atoms get together for a party in your wires, the outer electron just floats around in the metal. It always stays near somebody's nucleus, to cancel out the proton charge, but it doesn't really care whose nucleus.

The loose electrons become the *electron sea*, unbound and easily shifted around in the metal. All metals have this electron sea. Materials with loosely bound electrons are *conductors*. This doesn't mean, however, that you can pick up a wire and just shake the electrons off of it. The protons still keep their grip!

Most atoms are more jealous than metals and keep a tight grip on all of their electrons. When electrons can't move easily through a material, that material is an *insulator*. This is another reason why electrons don't jump off their wire. Where would they go? Attach themselves to another element, like the oxygen or nitrogen in the air? These elements don't want another electron, they have plenty already. So the electrons stay in place. They are insulated, blocked, prevented from moving off the metal.

In the real world most materials are not made from one pure element, and most elements are not perfect conductors or perfect insulators. Because of this variety, we can create a wide range of electron behaviors.

Electromagnetic Field

The electron has a negative charge and the proton has a positive charge. These charges affect the space around them, not unlike gravity, creating an *electric field*, also known as an *E-field*. Electrons and protons, together, are known as *charged particles*.

When you put a bunch of mass together in one place, you get a noticeable gravitational field. When you place an object on a tall hill in the presence of gravity, that field pulls it down to the bottom of the hill. When you take two large balls of mass, such as planets, they create such a large distortion that they are attracted toward each other across a huge expanse of space.

Both gravity and the electromagnetic forces reach out forever. The catch is, they get weaker with the square of the distance. At one "step" from the source, they have a strength of 1. Double that distance and they are 1/4 as strong, to 1/9, 1/16, and so on.

We deal with gravity every day and have a good sense of how it controls our lives. One aspect of gravity we don't notice is how quickly it "moves." When you wiggle a planet—okay, *if* you could wiggle a planet—the effect of this movement is that its gravitational field travels, or *propagates*, at the speed of light. So if the Earth was suddenly shifted a few thousands miles to the left, the moon wouldn't notice until the change in the gravity field reached it. Electric fields also propagate at the speed of light.

Gravity only attracts, while electric fields can attract or repel. It is as if the negative charge is a deep hole in the fabric of space and the positive charge is a tall hill. Hills push away from other hills, and holes push away other holes.

But hills and holes draw towards each other and cancel out, making space smooth again.

The electric field is different from gravity in another important way. It is much stronger, about 10^{36} times stronger. That's a 1 with 36 zeros after it, which is a really big number. Two grains of sand held close together have no noticeable gravitational attraction. If, however, the gravitational field was as strong as the electrical field, they would slam together with a force of over three tons.

The electric field is strong. A modest collection of electrons or protons, without their balancing partners, can create a huge electrical field. When you rub a balloon with a piece of fur, you create a modest electric charge on the balloon. Yet the electric field has enough strength to make your hairs stand on end from inches away.

When a charged particle is accelerated it creates a *magnetic field*, also known as a *B-field*. Both the electric field and the magnetic field are *vector fields*. A vector field has not only a strength but a direction. The electric field around an electron, for example, pulls protons toward the electron and pushes other electrons away.

The magnetic field created by a moving charge curls around the direction of motion. It is perpendicular to the direction of motion, but its idea of "perpendicular" is to wrap itself around the wire (Fig. 4-1). Note that a proton moving in the same direction as an electron will create a magnetic field of an opposite polarity.

A magnetic field that is changing in intensity generates an electric field. A changing electric field in turn generates a magnetic field. In fact, the total amount of field is constant, just rotating between magnetic and electric

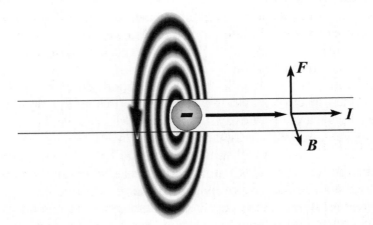

Fig. 4-1.   Magnetic field from electron motion.

forms. These two fields are so closely related that we talk about them as a single *electromagnetic field*.

When you accelerate a charged particle, a self-sustaining electromagnetic wave radiates out away from the motion. If you move the charged particle back and forth, you get a stream of these waves. Depending on how fast you move the charge, these waves could be radio waves, microwaves, and so forth.

If you move a wire through a magnetic field, or move a magnet past a wire, the moving magnetic field accelerates the charged particles in the wire. Even if the magnetic field is just getting stronger or weaker, it accelerates the electrons.

This is how radios work. On one side, the transmitter is moving electrons to send electromagnetic waves out into space. Somewhere else, these waves cross the wire antenna of a receiver and make its electrons move.

It is interesting to note that if you move the electrons in a wire to create a magnetic field, and then *stop* moving the electrons, the field goes away. While the field is collapsing, it is actually in motion and tries to push the electrons back in the direction they came from.

There is another piece to the electromagnetic puzzle, and this is the motion of the wire in relation to the magnetic field. If you hold a wire steady and move the magnetic field past the wire, the electrons are given a boost. This is how generators work.

Looking at this from the other direction, if you move the wire's electrons in the presence of a magnetic field, it generates a mechanical force between the electrons and the field. This is how motors work.

All three forces work at right angles to each other—the magnetic field vector B, the direction of the electrons moving in the wire I, and the force vector F pushing the wire. These are shown in Fig. 4-2. If you reverse any one of the vectors, one of the others must change to match. For example, keeping the B field constant and reversing I will cause F to reverse.

Units

UNIT PREFIXES

In mechanics, we worked with a limited scale of values. In electronics, however, we shall be working with both very large and very small values. The SI measurement system defines prefixes to scale values up and down in steps. See Table 4-1 for a list of these prefix modifiers. For example, 1,000 meters is a kilometer, or 1 km. 1/1000 of a meter is a millimeter, or 1 mm.

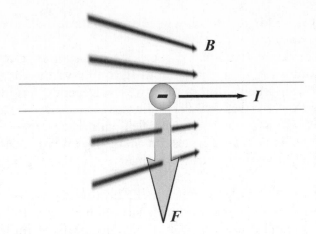

Fig. 4-2. Magnetic field, electron motion, and force vector.

Table 4-1 SI prefixes

Name	Prefix	Scale	
tera	T	1,000,000,000,000	10^{12}
giga	G	1,000,000,000	10^{9}
mega	M	1,000,000	10^{6}
kilo	K	1,000	10^{3}
milli	m	0.001	10^{-3}
micro	μ	0.000,001	10^{-6}
nano	n	0.000,000,001	10^{-9}
pico	p	0.000,000,000,001	10^{-12}

ELECTRICAL CHARGE

Coulomb: C

The *Coulomb* is a measurement of electrical charge. One coulomb is equal to the charge of 6.24×10^{18} electrons, which is a lot of electrons.

CURRENT: *I*

Coulombs per second: $I = C/s$

The flow of charge past a point in an electrical circuit is known as *current* and is represented by the symbol *I*.

The *ampere* or *amp*, symbol *A*, is the unit that current is measured in. One amp of current is one coulomb of charge passing by a point in one second. If you watch your circuit for one second and 6.24×10^{18} electrons march by, that's one amp.

I've used the word *circuit* twice now. Electricity is useful when it is moving from one point to another, doing work as it goes. The path that electricity takes is called a circuit. A nonelectrical definition of "circuit" is that of a path going in a circle, such as a race track.

Since it's not feasible to count the individual electrons flowing in a circuit, we usually measure the amp by other means. The amp can be measured because of the electromagnetic fields generated by electrical current. We measure this field and, from this, get a measurement of current. While the coulomb is a fundamental value, the *measurement* of the coulomb is based on the measurement of the amp.

Note that, in electronics, the electrons move from the negative terminal in a battery or generator to the positive terminal. This is *electron current*. In many discussions of electronic circuits, the convention is to imagine that the current flows from the positive terminal to the negative, opposite of what actually happens.

This *conventional current* is based on the history of Benjamin Franklin's observations of electricity. As more of the true details of electricity were discovered, the convention of current flowing from positive to negative remained.

This doesn't actually affect anything. If the protons really were flowing from the positive terminal of the battery to the negative, their effect would be exactly what we see for the electron flow we do have. The magnetic fields are the same and the mechanical forces applied to the wire are the same.

When we reverse the motion of a charged particle, all of the vector fields associated with it reverse as well. When we reverse the polarity of a charge particle, this reverses the polarity of the fields. If you reverse both the direction and the polarity, the associated fields remain unchanged. So electrons flowing in one direction have the exact same effect as protons moving in the opposite direction.

CHARGE DIFFERENCE

Volt: *V*

Voltage may be represented by the symbol *V* or *E*. In this book we use *V* exclusively.

Voltage does not exist at a single point in a circuit, but is a measurement of the difference in electric potential between two points.

Remember the potential energy we discussed in Chapter 2? The potential energy of an object was defined in relation to its height above ground level. Voltage measurements have the same need for a "ground" or reference point to measure from.

Voltage is based on *electrical charge*. Say you have a battery or generator that is pumping electrons into one end of a wire. Electrons find each other repulsive, so they try to stay as far away from each other as they can. But what if the wire doesn't connect to the other side of the battery, so the electrons in the wire have no place to escape to? They crowd closer together, but they don't enjoy it. The tighter they squeeze, the more they push against each other. Like springs, they are storing energy as they are squeezed.

This creates a *charge imbalance*, where part of the circuit has an unusual quantity of electrons. In turn, the force created by this imbalance is known as the *electromotive force* (EMF).

Voltage is the measurement of the difference in pressure from one point in a circuit to another. If the whole circuit is at the same pressure, the voltage will be zero even if it holds a hefty electric charge.

Water provides a popular analogy for electricity. The water molecules are like the electrons, so a gallon of water is like a coulomb of electricity. Water pipes are like wires, and the flow of water stands in for electric current. Voltage is represented by water pressure. Gravity is usually used for the electric fields.

The water analogy gives us illustrations like Fig. 4-3, where two tanks have different water levels and the pressure of water trying to flow from the left tank to the right tank is the voltage between the tanks. If the valve were opened, the current would be the flow through the valve.

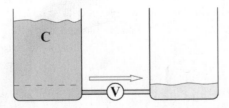

Fig. 4-3.   Water analogy.

ELECTRICAL ENERGY

Joule: $J = V \times C$

The unit of energy, the familiar joule, is also used for electrical energy. The joule provides a crossover point between mechanical and electrical measurements.

One joule is the energy needed to move one coulomb of charge up one volt of electrical potential. It is also defined as the amount of work performed by a current of one amp acting against a resistance of one ohm for one second. Resistance is discussed in the next chapter.

POWER: *P*

Watt: $W = V \times I$

Work or power is represented by the *Watt*, which is a function of voltage and current. The watt could also be considered joules per second, $W = J/s$.

Batteries and Generators

This discussion of electricity has focused on the flow (current) and pressure (voltage) of the electric charges. The question is then, how do we create this voltage? With electron pumps.

The electricity that comes out of the walls in your house is pumped there by giant generators. These generators use magnets to push the electrons around.

Batteries do the same thing but on a smaller scale and without magnets. Inside the battery is a chemical paste that acts as a pump. As the chemical reaction inside the battery goes forwards, it charges its terminals. Batteries are like chemical springs unwinding, converting chemical potential energy into electrical energy. When you recharge a battery, you are rewinding this chemical spring.

Inside a battery, a chemical *redox* (reduction/oxidation) reaction occurs at two electrodes, or conductive posts, which are separated by a conductive fluid or paste, the *electrolyte*. The reduction reaction occurs at the *cathode*, which is the positive electrode. The oxidation reaction occurs at the *anode*, which is the negative electrode.

As the battery discharges, powering your circuit, the anode oxidizes (picks up oxygen) and creates a surplus of electrons. These electrons then flow through your circuit (the load) and reenter the battery at the cathode. They can get back in through the cathode because it is performing the reduction reaction (losing oxygen), which combines the electrons with the cathode. The electrolyte that bathes both the anode and cathode finishes this electron loop, allowing the electrons to hitch a ride to the anode for another trip through the oxidation reaction. Note that batteries don't create electrons; they just provide a way to move them around.

Of course it's far more complicated than that. It's not just electrons, the negatively charged particles, that move. Protons also flow in the electrolyte, and the chemical reactions are fairly complicated.

Electrons flow from the anode to the cathode in the metal wires of an electrical circuit. In the electrolyte of a battery there is a flow, in opposite directions, of both positively and negatively charged atoms, or *ions*. The ion flow inside the electrolyte must exactly match the electron flow outside the battery through the load. In fact, the internal resistance of the battery to the ionic flow is an important factor in how much energy the battery can provide at a given time.

The terms "anode" and "cathode" can be confusing, because they seem to mean different things when applied to batteries versus other components. The cathode is the negative side of a two-wire component such as a resistor, but it is the positive terminal of the battery. Both definitions are correct.

The anode is, formally, the terminal or electrode where electrons leave a system. In a current source like a battery or generator the anode is the negative terminal because electrons are being pushed out from it. But in an energy-consuming component it is the positive terminal, since the electrons are being pulled out of the component by the positive charge on that side. Conversely, the cathode is positive on a battery yet negative on a passive component.

When electrons leave a battery they enter a component—so the anode (electron source) of the battery is connected to the cathode (electron sink) of a component.

SPEED OF ELECTRICITY

Electric charges move very slowly through your wires, slow like the minute hand on a clock is slow. And yet, when you turn on a light switch, the power reaches the bulb at, roughly, the speed of light.

The electrons in your wire don't move fast, but they move together. Here is another electrical analogy, the clothesline. Take two pulleys and anchor them on two different sides of your yard. Wrap a rope into a circle, through these pulleys. When you pull the rope on one side of the yard, the pulley on the other side feels the force "instantly." The rope itself is moving slowly, but the power is transmitted by the rope quickly.

Summary

This chapter covered a lot of territory, nothing less than the structure of everything in the universe!

We looked the closest at the electron and the electric and magnetic fields associated with it. Because of the special relationship between magnetism and electric charges, we can use magnets to pump electricity. Using these same fields in the other direction, we can use electricity to push against the magnets.

Quiz

1. What is the difference between a conductor and an insulator?
2. What is electricity?
3. How do the magnetic field and the electric field interact? How does this relate to the way radios work? Electric generators? Electric motors?
4. What is a circuit?

I'm not going to ask you to define amps (coulombs of charge moving past a point per second) or voltage (charge difference) because, frankly, you can look up the formulas. It is more important to have an understanding of what electricity does before you get buried in the math describing it. Of course, in later quizzes I may not be so easy on you, so keep an eye on those equations.

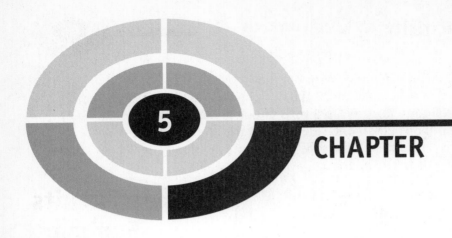

CHAPTER

Starting with Electronics

Introduction

Enough with the theory! The terms and definitions and forces are important, since they give us the language and concepts we need to do our work, but learning the nature of the joule isn't where the fun is. This chapter is where we start to put theory into practice.

First, of course, there is a bit more theory to learn. To describe electronic circuits, we draw them using circuit diagrams, or schematics, so you need to learn how to read those.

Then there are a few practical details of where to get electronic components and how to wire them together.

Then we get to play with that most basic of all components, the resistor. Even though a resistor doesn't "do" anything, except perhaps turn electricity into heat, it has several subtle uses. It also provides illustrations for some of the forces we talked about in Chapter 4.

Electronic Circuits

SCHEMATIC

A *schematic* is a diagram that represents, in abstract lines and squiggles, an electronic circuit. It is almost like a road map, but even more abstract since a road map shows where things are positioned in space relative to each other and a schematic does not. The schematic only shows how the *components*, or parts in an electronic circuit, are connected to each other.

An example of a schematic for a simple circuit is shown in Fig. 5-1. There are several features in this figure that are common to all schematics.

The little pads on the left labeled "+5" and "Gnd" are plugs to the outside world. External to this circuit we expect there to be a 5-volt power supply. Electricity is always measured with respect to a *ground* state. It's handy to label which wires we expect to be at zero volts, or ground.

The inverted pyramid in the bottom-left corner is one of several symbols that mean "ground." One version of ground has only one horizontal line instead of a pyramid of them. Note that ground is our source of electrons.

The lines in the schematic are the wires that connect everything together. A round dot at a junction of lines means that these wires are connected. Sometimes you have to draw wires in the schematic so they cross each other

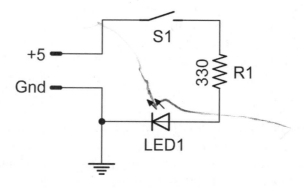

Fig. 5-1. Schematic.

without connecting. Crossed lines will not have a dot, or may even have a little arch in them so that they obviously cross.

Each component in a schematic has its own symbol. Note that European and American schematics tend to use different symbols. We focus on the American symbols.

The symbol above the label "LED1" is for a light-emitting diode. Component labels tend to follow a simple convention. LEDs are "LED," resistors are named "R," capacitors "C," switches "S," and so forth. The number after the name is just a counter, 1, 2, 3, to keep the various parts separate.

Schematics usually come accompanied by a parts list that gives values for all of the names. The values may also be printed on the schematic, like the 330-ohm resistor on the right.

Switches and relays, which are just electrically triggered switches, are normally drawn in their unactuated or default position. As we introduce various components in the projects, we shall show their schematic symbol and name, as well as what they look like physically.

PRINTED CIRCUIT BOARD

If a circuit is going to be of any use, the lines and components in the schematic need to be translated into physical reality. Most electronic circuits that you find in the wild are assembled on *printed circuit boards* (PCBs), or just *circuit boards*. The circuit board for our schematic is shown in Fig. 5-2.

Most PCBs are created photographically. The special fiber board begins its life covered on one or both sides with copper. A photoresist mask is applied to this and then a picture of the circuit is projected onto it. Most of the mask is then washed off, leaving mask over just the traces. A strong acid is used to dissolve the rest of the copper off of the board. When that is done, holes can be drilled for the components and the board is done.

The *traces* on the board are wires. Flat wires, but still the same thing as the wires you are used to. The *pads* are round or oval areas where components are inserted into the board and soldered into place. The process of soldering is described later.

Cheap circuit boards have copper on just one side, or maybe on both sides, but no copper in the holes. Better circuit boards have *plated holes*, where copper is plated into the holes on the board. This copper plating connects to the solder pad and makes it easier to solder components into the board.

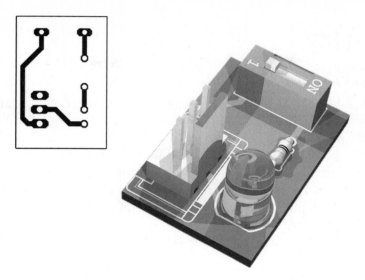

Fig. 5-2. Circuit board.

Plated holes are also stronger and the pads are less likely to "lift" off the board. The copper is held onto the board with glue, but too much heat during soldering can melt the glue and cause the pad and traces to lose their grip on the board.

If there is copper on both sides of the board, plated holes also do a good job of connecting the pads on the top and bottom sides of the board. There may even be plated holes in the board that aren't for components, used to connect traces between the front and back. These are called *vias*.

Some circuit boards have more than two layers of copper. These multilayer boards are like several two-layer circuit boards glued, or laminated, together. In these cases the vias are critical for connections between one layer and the next.

While a professional circuit board is created using computer-aided design (CAD) software, you can also make them by hand using either a chemical photoresist or stick-on resist pads and traces. I don't personally recommend these kits, since good results can be hard to achieve and the chemicals are particularly nasty.

To do your own professional schematic layout and circuit board design, I recommend the freeware version of Eagle, from CADSoft. For not too much money, you can even find companies that specialize in creating proto-type circuit boards. My favorite prototype board manufacturer is Alberta Printed Circuits in Canada, but there are others listed later which also do good work.

Circuit Assembly

PROTOTYPING BOARDS

When you are experimenting with a new circuit, you normally want to put it together in a temporary fashion to test it. Test versions of a circuit are *prototypes*, and prototyping boards, often called *breadboards*, make it easy to quickly wire a circuit. An example breadboard is shown in Fig. 5-3.

Once a circuit has been tested, you may want to make a permanent version using either a generic or custom circuit board. Each hole in the breadboard is a socket that you plug wires or components into. The holes are connected together in a standard pattern, as shown in Fig. 5-4.

The long *bus* or *rail,* a series of connected holes on the outside edges of the breadboard, is normally wired to the power supply. I usually take a red and black permanent marker and draw lines along these holes, to identify which strip is which. Black is the color traditionally used to represent ground and

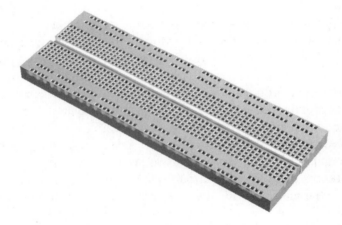

Fig. 5-3.   Prototyping breadboard.

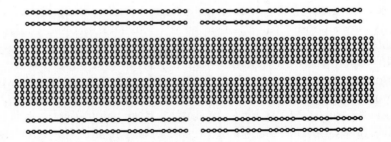

Fig. 5-4.   Breadboard circuit layout.

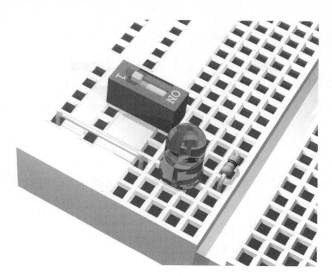

Fig. 5-5. Breadboarded circuit.

red is positive voltage. On a long breadboard like this one, the power rails tend to be split in the middle. I'll also usually take short wires and connect them together into one long bus. Our previous circuit, plugged into the breadboard, might look like Fig. 5-5.

Breadboards come in many different sizes, though the internal wiring is usually pretty standard. The 64-unit long board shown here is a common size. Other breadboards may come with interlocking wedges along the sides so you can connect several boards together into one unit.

You can also find prototyping stations that come with multiple breadboards plus built-in power supplies and other attachments. These cost more, but can make the task of prototyping easier.

You can buy circuit boards that are laid out like the breadboards. These let you solder your components and jumper wires into place permanently. You can often take a design straight from the breadboard onto the matching circuit board.

DEAD BUG AND WIRE WRAPPING

There is one inelegant method of assembling components that is often referred to as the *dead bug* technique. It is called this because it is normally used with integrated circuits, which look like futuristic bugs.

Dead bug assembly is where you solder the wires and components together without any type of printed circuit board, prototyping board, or other structural support. The circuit simply hangs together in space, waiting to get bumped and shorted out.

Wire wrapping replaces soldering with tightly wrapped wires. Though a popular prototyping method, I've never enjoyed the process so can't recommend it.

SOLDERING

Most of the work in this book is done on breadboards, so soldering is a useful skill to have. *Soldering* is the process of joining two wires together, or a wire to a circuit board, by melting a soft metal, *solder*, into the junction.

Equipment

Solder (Fig. 5-6) is a soft metal wire made from a mix of tin (symbol Sn) and lead (symbol Pb), usually about 60% tin and 40% lead. My solder is marked *SN63PB37*, indicating the popular mixture of 63% tin and 37% lead by weight. This solder melts at the low temperature, for metals, of 190°C (374°F).

Since there is lead in solder, it is a somewhat hazardous material. You don't want to eat lead or rub it in your eyes, so wash your hands after handling solder.

When metals get hot they react to the oxygen in the air more than they would normally. This creates an oxidized layer on the metal that interferes with the solder sticking. It also insulates the metal from what we are trying to solder it to.

Fig. 5-6. Solder.

To prevent oxidation, as well as to clean off existing oxidation, you use solder with a *flux*. In plumbing and other large-scale soldering projects this flux is an acid paste. Never use acid flux in electronics! It will damage your circuit.

Electronic flux is usually a rosin paste, based on pine-tree sap. For convenience, most electronic solders are formed as hollow tubes with a rosin flux core. Sometimes you want lots of extra flux, so you can buy rosin flux in liquid form, thinned with alcohol.

Solder is heated with a *soldering iron*, such as my old workhorse shown in Fig. 5-7. A basic soldering iron is a cool handle with a hot tip, plus some kind of stand to keep it from burning your desk. A better soldering iron includes temperature controls and a place for your sponge. All irons feature replaceable tips, since these wear out over time.

Since you have a choice of tip, use a small pointy tip. While you can get large tips, flat screwdriver-shaped tips, and so on, these are clumsy for most electronics work. It is important to keep your tip clean. This is what the sponge is for. Never use sandpaper or other abrasive materials to clean your tip, since these will damage its protective plating.

When your soldering iron is hot, you clean it by wiping the tip on a damp sponge. Once you wipe off as much crud as you can that way, cover the hot

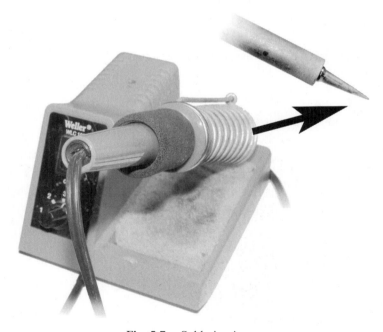

Fig. 5-7.   Soldering iron.

Fig. 5-8.   Third hand.

tip with solder. The rosin flux will help clean the tip further, and the coating of solder keeps the tip from oxidizing again.

This process of cleaning and coating the tip of your soldering iron is called *tinning*. You may need to tin and wipe the tip several times before it becomes clean. You can tell when the tip is properly tinned when it shines a bright silver, rather than a dull gray.

The wire or component leads also need to be clean. If they are dull and dirty, you can scrape them clean with a sharp blade (which will soon be made dull) or sandpaper. Dirt, oil, or oxidations keep you from getting a good solder joint. So keep everything clean.

You can tin the ends of your wires, and this can make a big difference in how easy they are to solder. Note, however, that tinned stranded wire is hard to bend, so you want to shape it first.

When soldering wires, it can be handy to have a third hand (Fig. 5-8) to hold the components. Your other two hands are holding the iron and the solder, respectively. Hobby stores and many electronics suppliers provide various clamps and clips designed to act as a third hand. Some of these are specially designed to hold circuit boards.

Wire-to-wire connection

Try This: There are times when you need to solder two wires together. This section goes over the soldering process.

First a word about wires. There are two basic types of wire, solid and stranded. Solid wire is just that, a single solid wire, usually inside some type of plastic insulation. The is your standard "jumper wire" used to connect components on breadboards and circuit boards. When you work with solid wire you need to be careful not to nick it, or it is likely to break at that point.

Stranded wire comes in many varieties and forms, but has the common trait of being made up of many little wires twisted or braided into a larger wire. Stranded wire doesn't work well for board-to-board connections, since the little strands tend to get rowdy and provide short circuits. If you have a job where the wire might bend during use, such as attaching a sensor or control, or are carrying a signal a long way from the circuit, you need to use stranded wire.

A *short circuit* is where you create an accidental connection where you don't want one. It creates a shortcut through your circuit, and is almost always going to prevent the circuit from working correctly.

Stranded wire holds up to being bent and flexed a lot better than solid wire. So, solid wire for nonmoving jobs, stranded wire for things that flex.

The first step in using wire is to cut it to length. The second step is to strip the ends so a short section of bare wire is exposed (Fig. 5-9).

Solder is not very strong, so you need to make a mechanical connection between your wires. I like to use round-nosed pliers to make a loop in the end of the wire (Fig. 5-10), though you can use any type of small plier to bend the wire.

While the loop-to-loop connection works well for component leads, you can also twist the ends of wires together to make a splice. There are many different ways to connect wires together, so feel free to experiment to get the best connection for your projects.

Holding the wire in a third hand, or placing it on a heatproof surface, you want to touch the end of your soldering iron to the junction of the two wires. You need to heat both wires to the solder's melting temperature to get a good joint.

I prefer to have my soldering iron heavily tinned, so there is actually a tiny drop of solder on the tip. This gives a better path for the heat to flow from the iron to the wire than a bare tip. Once the wires are hot, touch the solder to

Fig. 5-9.   Stripped wire.

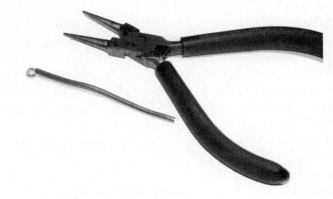

Fig. 5-10.   Bend wire.

Fig. 5-11.   Solder wire.

them until the solder melts into a small, shiny blob. Be careful not to move the wires while the solder is cooling, or it can make a poor connection. Also be sure you aren't applying the solder to the soldering iron. The melted metal flows to where the heat is, and if you touch it to the hottest part, the iron, it won't flow around the wires where you want it. See Fig. 5-11.

Wire to board

Soldering a wire or component to a circuit board is easier than wire-to-wire. The process is essentially the same as wire-to-wire soldering, only you insert your wire into a hole in the circuit board to start. You can bend the wire or component lead a little bit so it doesn't fall out when you turn the board over to solder it.

Figure 5-12 shows the component in the circuit board with the wires poking out through the back of the board. The soldering iron is heating the wire and the pad together. Once you have soldered the wire to the pad, you will have a shiny cone that flows around the entire pad and up the wire. A "blobby" junction needs to be reheated, taking care to heat both the wire and the pad.

If the holes in your board are not plated through, it can be difficult to get a good solder junction. You may need to put in extra solder to bridge the gap from the pad to the wire. Be careful, though, since too much solder can create a *solder bridge* to another pad and create a short circuit.

Once soldered into place, trim the wires down to the top of the solder cone (Fig. 5-13) and turn the board over to admire your work (Fig. 5-14). I can almost never get my parts to sit flat on the board, but minor aesthetic details like that don't matter. As long as the solder joint is good, the part isn't too high off the board, and you didn't create a short, it's good. If the

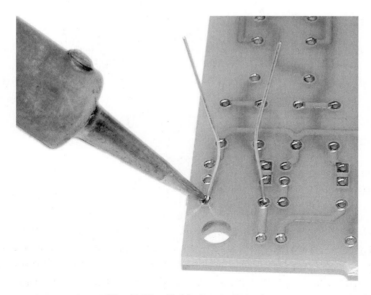

Fig. 5-12. Soldering to board.

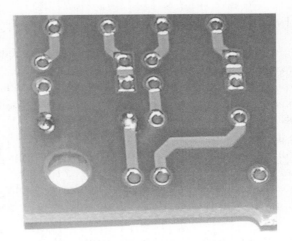

Fig. 5-13. Trim wires.

Fig. 5-14. Resistor on board.

part is too far off the board, it could bend over and short against a neighboring part.

Desoldering

Since nobody is perfect, you may have to remove the solder from a joint. There are different ways to do this.

One way is with desoldering braid, which is a type of stranded wire. You lay the braid over the soldered joint and heat it with the soldering iron. Once the solder flows into the braid you can remove it, the iron, and the solder from the joint.

You can also get different types of "solder suckers." These can be small squeeze bulbs or spring-action devices that provide a quick inhalation. The idea is to heat up the solder so it is liquid and then suck it into the device. These actually work, and high-end soldering stations include a small vacuum-pump desoldering nozzle. Either way, you may need to heat up the wire and pad again to be able to pull the wire out of its hole, since there will still be a small amount of solder "gluing" it together.

As with all things, soldering and desoldering are skills that will get better as you practice. In fact, you may want to practice now, before you need to solder something you care about.

SUPPLIERS

To work with electronics, you need to find a source for parts. Radio Shack has a fair selection of parts, and they should have everything that we use in this book. For many people, they are the most convenient source of parts. Fry's Electronics, when you have one in your neighborhood, is also a good source of parts. Most larger towns also have some kind of electronic supply store that supports the local electronics repair and hobbyist communities. You can find these using your local Yellow Pages.

Almost everyone has access to mail order suppliers. Table 5-1 lists a number of the more common mail order electronics suppliers. If you want to branch out into circuit board design, Table 5-2 lists a few manufacturers of prototype boards. Note that CadSoft doesn't make circuit boards, but has a freeware software package you can use to design them.

Table 5-1   Electronics suppliers

All Electronics	http://www.allcorp.com/
Digi-Key Corporation	http://www.digikey.com/
Fry's Electronics	http://www.outpost.com/
Jameco Electronics	http://www.jameco.com/
JDR Microdevices	http://www.jdr.com/
Mouser	http://www.mouser.com/
Radio Shack	http://www.radioshack.com/

Table 5-2   Prototype circuit board manufacturers

AP Circuits	http://www.apcircuits.com/
CadSoft	http://www.cadsoft.de/
ExpressPCB	http://www.expresspcb.com/
PCB Express	http://www.pcbexpress.com/

Resistors

RESISTOR

Everything acts as a resistor to some extent. However, there is also a specific electronic component called the *resistor* that provides a calibrated resistance. Resistance is measured in *ohms*, and the ohm is represented by the Greek letter Ω (omega).

Resistance is the measure of a material's opposition to the movement of electrical charge. As you are trying to pump electrons through a circuit, the resistance of the circuit is fighting back, keeping the electrons from moving as easily as they might.

Everything has some resistance, even the wire and copper traces in a circuit board that you use to connect components together. While many resistors are made from a carbon film, like a conductive paint, highly accurate resistors are made from fine wire. Usually, though, the resistance in your wire is too small to make any difference for low-power circuits.

One of the byproducts of resistance is heat. The harder you push electrons through a resistor, the more heat is created. The energy you are using to push electrons through the resistor doesn't make it all the way through, since the resistor is opposing that force. The result, then, is heat. Using the clothesline analogy again, as you are pulling on the clothesline, someone in the middle is pinching the rope with their fingers. This slows down the rope, or at least makes it harder to pull, at the expense of making their fingers hot.

Looking at Fig. 5-15, you can see the schematic symbol for the resistor is a jagged line. The resistor itself is a cylindrical blob on a wire, with three or four stripes. The stripes will normally be closer to one end of the resistor than the other. To read the stripes, orient the resistor so the stripes are on the left. Note that resistors don't care which direction you plug them into circuit. We put the stripes on the left for easy reading.

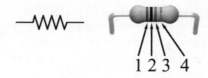

Fig. 5-15. Resistor.

The first and second stripes are color codes for different digits. Black is 0, Brown is 1, and so forth. All of the colors and their meanings are listed in Table 5-3. The third stripe is a multiplier, and it determines how many zeros you add after the first two digits. A resistor marked Black, Brown, Black has a value of 1 Ω. Black, Brown, Brown adds a zero, so the value is 10 Ω, and so forth.

The reverse is also true. When you see a resistor valued at 4.7K on a schematic, you want a resistor whose first two stripes are Yellow and Violet. Since "K" means 1,000, the full value is 4,700. Since we need to add two zeros, the third stripe is Red.

The fourth and last stripe is an optional *tolerance* indicator. No device is perfect, so this stripe indicates how imperfect the resistor is. If there is no fourth band, the resistor's actual value is plus or minus 20%, or somewhere between $(R \times 0.80)$ and $(R \times 1.20)$. Better resistors have tighter tolerances, as indicated in Table 5-3.

In addition to the resistance value, resistors also have a power rating. The power rating determines how much power, in the form of heat, the resistor can handle. Most circuits use 1/8 or 1/4 watt resistors. When a larger resistor is needed, the power rating is usually specified in the parts list.

OHM'S LAW

The unit of resistance, the ohm, is named after the German physicist Georg Ohm. His work had a strong influence on our understanding of electricity and resistance. The relationship of resistance to current and voltage is known as *Ohm's Law* in his honor.

Ohm's Law states:

$$V = I \times R \tag{5-1}$$

Equation (5-1) states that when a current encounters friction in the form of resistance, a voltage appears across the resistor. Remember that voltage is like a pressure difference between two points in the circuit. When the

Table 5-3   Resistor color code

Color	Digit	Multiplier
Black	0	1
Brown	1	10
Red	2	100
Orange	3	1,000
Yellow	4	10,000
Green	5	100,000
Blue	6	1,000,000
Violet	7	10,000,000
Gray	8	100,000,000
White	9	not used
	Tolerance	
(no stripe)	± 20%	
Silver	± 10%	
Gold	± 5%	
Brown	± 1%	

electrons meet resistance, they don't flow as easily through it. This creates a reduced electrical "pressure" downstream from the resistor.

The more resistance given to a current, the higher the voltage difference. The traditional description of Ohm's Law is that a potential difference of 1 volt will push a current of 1 amp through 1 ohm of resistance.

Different permutations of this equation give different ways of looking at the circuit.

$$I = \frac{V}{R} \tag{5-2}$$

Equation (5-2) says that for a given voltage, the smaller the resistance the more current can flow through it. Or it could say that for a given resistor, a higher voltage can push more current through it.

$$R = \frac{V}{I} \tag{5-3}$$

Equation (5-3) is the third form, for completeness. It shows how you calculate resistance using the voltage across the resistor and a known current flow.

RESISTOR NETWORKS

It is a rare circuit that has just one resistor in it. You will usually have several resistors in different configurations in the circuit. How do you calculate the overall resistance of these circuits?

First, why would you want to calculate the resistance of a resistor network? Because if you know the resistance of a circuit and the voltage it is operating at, both of which are usually known values, you can calculate the current consumed by the circuit using equation (5-2). Knowing how much current a circuit needs, you can decide what kind of power supply it needs as well as how much power, typically released as heat, it is going to use.

The power consumption of a circuit, in watts, is determined by equation (5-4). For the second form of the equation, we replace V with its Ohm's Law equivalent $I \times R$:

$$P = I \times V$$

$$P = I^2 \times R \tag{5-4}$$

A small difference in current can make a big difference in power consumption.

The calculation of resistance depends on how the resistors are wired together.

Series network

When devices are in *series*, it means that they are connected end-to-end, as shown in Fig. 5-16. Resistors in series add together to make a

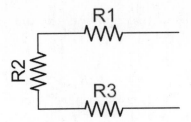

Fig. 5-16.   Series resistors.

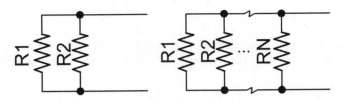

Fig. 5-17.   Parallel resistors.

larger resistor:

$$R = R1 + R2 + R3 \tag{5-5}$$

Parallel network

When devices are in *parallel*, they are connected side-by-side, as shown in Fig. 5-17. Because current has more than one path through the circuit, there is less resistance in a parallel circuit than in one of the resistors in the circuit. For two resistors, the total resistance is:

$$R = \frac{R1 \times R2}{R1 + R2} \tag{5-6}$$

If there are more than two resistors, the general calculation is:

$$R = \frac{1}{\dfrac{1}{R1} + \dfrac{1}{R2} \cdots \dfrac{1}{RN}} \tag{5-7}$$

Voltage divider

One application of a resistor network is as a *voltage divider*, shown in Fig. 5-18. When a voltage is applied across this network, the output voltage

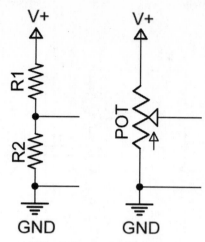

Fig. 5-18.   Voltage divider.

at the point between the two resistors is a controlled fraction of the input voltage:

$$V_{\text{out}} = V_{\text{in}} \times \left(\frac{R2}{R1 + R2} \right) \tag{5-8}$$

If one of the resistors, such as *R2*, isn't fixed but is allowed to change in value, the output voltage will change from some high value to nearly zero depending on the setting. This can then be used as a control value, such as a volume control, or provide the result from an input sensor.

A *potentiometer*, or variable resistor, is often used as a resistor bridge to provide control for electronic devices. On schematics these can be abbreviated *pot* or labeled *trim* for trim potentiometers that are used to adjust a circuit.

Wheatstone Bridge

The resistor network known as the Wheatstone Bridge is named after Sir Charles Wheatstone, who found many uses for it in and around 1843. He didn't invent, it, though. Samuel Hunter Christie documented this bridge in 1833.

The Wheatstone Bridge consists of two voltage dividers, as shown in Fig. 5-19. Typically, one of the dividers has a fixed value and provides a reference voltage. The other divider has a variable resistor in it, such as *R2*.

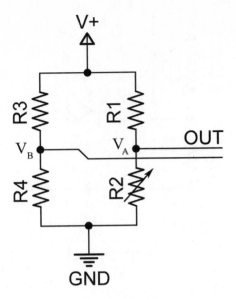

Fig. 5-19. Wheatstone Bridge.

This may be a potentiometer or, more likely, some kind of sensor like a photoresistor, strain gauge, or thermistor (a typical temperature sensor). The output voltage is the difference of the output of the two voltage dividers:

$$V_{out} = V_A - V_B$$

Expanding from equation (5-8), we get:

$$V_{out} = V_{in} \times \left(\frac{R2}{R1 + R2} - \frac{R4}{R3 + R4} \right) \qquad (5\text{-}9)$$

Another use for the Wheatstone Bridge is to measure an unknown resistance. If, for example, $R3$ was an unknown resistor, we can adjust the calibrated $R2$ until the output voltage goes to zero. $R3$ is then equal to:

$$\frac{R3}{R4} = \frac{R1}{R2}$$

$$R3 = R4 \times \frac{R1}{R2} \qquad (5\text{-}10)$$

To get good results, all of the known resistors need to have tight tolerance values to reduce the amount of error in the circuit.

RESISTIVE SENSOR

Try This: This little project will give you some hands-on experience with resistors, voltage, and the Wheatstone Bridge. You will need the various tools and parts listed in Table 5-4 to complete this project.

Components

There are a few components other than the $10\,k\Omega$ resistors. The first thing you need for any electronics project is electricity. There are many different ways to get power to your circuit. You can use a power adapter that you plug into the wall. These power adapters, sometimes called "wall warts," come in many different sizes, voltages, and current ratings.

A more portable solution is to wire a battery into the circuit. You can get battery packs for AA cells, C cells, and even clips for those small calculator and watch batteries. My preferred battery is the 9-volt cell. You can find clips, as shown in Fig. 5-20, for these 9-volt batteries. Make sure the ends of the wires are long enough and tinned so they are stiff enough to push into your breadboard. You may even want to solder a length of solid-core jumper wire to the clip, so it is easy to fit into your breadboard.

Table 5-4   Parts: resistive sensor

Parts	
R1, R3, R4	$10\,k\Omega$ resistors
R2	$100\,k\Omega$ photoresistor
R5	$10\,k\Omega$ potentiometer
J1	9-volt battery clip
BAT	9-volt battery
Tools	
Breadboard	
24 ga jumper wire	
Multimeter	
Small pliers	

Figure 5-21 shows two easy to find variable-resistance sensors. The part on the left is a cadmium sulfide (CdS) *photoresistor*. The part on the right is a *thermistor*. A photoresistor changes its resistance based on how much light is falling on it. They have typical resistances from 100 kΩ in the dark to about 10 kΩ in bright light. The thermistor changes resistance based on what temperature it is, from about 320 kΩ in an industrial deep-freeze to about 900 Ω in boiling water. We use the photoresistor in this project.

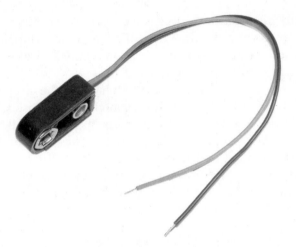

Fig. 5-20.   9-volt battery clip.

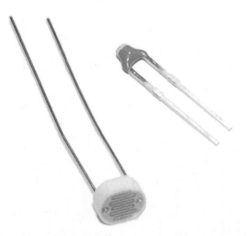

Fig. 5-21.   Photoresistor and thermistor.

Fig. 5-22.   Potentiometers.

We shall also need a $10\,k\Omega$ potentiometer for this project, to fine-tune the results. These take all sorts of shapes and forms, as shown in Fig. 5-22. The named resistance for a potentiometer is its maximum resistance, The actual resistance of the a third lead depends on the position of the potentiometer's knob. If you want the ability to really fine-tune your circuit, track down a multi-turn potentiometer. Normal potentiometers go from zero to full resistance in less than one turn. For finer control, you can find versions that take multiple full rotations, as many as ten or even twenty-five, to cover the same resistance range. These are more expensive but provide better control.

You will probably need to solder wires to your potentiometer, unless you can find a very small trim pot that will plug directly into the breadboard.

Tools

If you come across a component and you don't know what its resistance is, you can test it with an *ohmmeter*, which measures resistance, or a *multimeter* (Fig. 5-23) which measures ohms and several other types of values. Another function on the multimeter is the *voltmeter* to measure voltage.

These testing tools usually have a dial where you can select the test type and an operating range. For example, in the pictured multimeter, if I want to test what I think is a $100\,k\Omega$ photoresistor I would dial it to

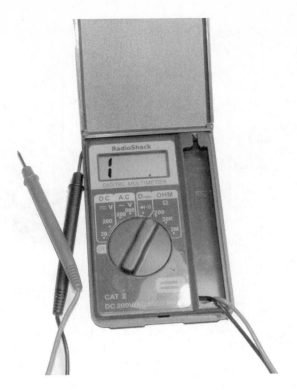

Fig. 5-23.   Multimeter.

the 2 MΩ range and apply the probes to the leads of the resistor. A result of 0.10 would be in the megaohm (0.1 MΩ) range, meaning 100 kΩ. More detailed information should come in your multimeter's instruction pamphlet.

Try this out on several known resistors to get a feel for the process. Note that if your hands touch both of the leads on the component you are testing, or the metals tips of both probes, the resistance of your body will be tested in parallel with that of the resistor. It can by tricky to test a part without actually touching it, but it is important that you do.

In addition to the multimeter, you should have at least one pair of small, needle-nosed or round-nosed pliers. The project itself is built on a breadboard, so it would help to have one of those too.

For jumper wires, you can get a spool of solid-core wire in the 22 to 24 gauge size. The wire's *gauge* (ga or AWG for American Wire Gauge) is a measure of how big it is. The larger the gauge the smaller the wire's diameter. 24 ga wire is about 0.20 inch in diameter. It is easier to buy a package of precut, stripped, and bent wires than to make your own little wires to connect the parts on the breadboard. These come in many handy sizes and are color-coded by length.

Project

Figure 5-24 shows the final schematic for this project. We don't build it all at once, however, but start with one voltage divider using *R*1 and *R*2. Note that this is a simple extension of Fig. 5-19.

The large arrow through *R*2 indicates that it has variable resistance. The two small arrows pointing into *R*2 indicate that it is sensitive to light. Together, these symbols tell us that this resistor changes its value based on the light hitting it.

Put together the circuit shown in Fig. 5-25, our first voltage bridge. Bend *R*1 so that it will fit from the positive voltage rail to an interior row. The photoresistor can then be plugged into the next row over and into the ground rail. A short jumper then connects the two breadboard rows. Take your 9-volt clip and plug the red (positive) lead into the positive rail, and the black lead into the other rail. Finally, attach the battery.

It can be difficult to test the voltage across different holes in a breadboard, since the probes tend not to fit into the holes. You can always touch the probes to the leads of components plugged into the board.

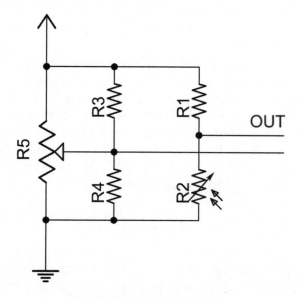

Fig. 5-24.   Final sensor circuit.

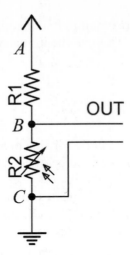

Fig. 5-25. Photoresistor voltage bridge.

If you test the voltage from the positive rail to the ground rail, you will get a value something like 9 V. It won't be exactly 9 V, since the voltage across batteries changes with time and their charge. If you get a negative voltage reading, make sure the black probe goes to the black, ground, wire.

With the circuit hooked up you can test it from two different perspectives. Try putting the positive probe on the 9 V rail (A) and the ground probe at the junction (B). When $R2$ is in bright light the AB voltage is nearly 9 V. In the dark, AB is about 4.5 V.

Looking at the circuit from the perspective of BC, with the positive probe at B, you get readings near zero for bright light and again about 4.5 V in the dark. The circuit was designed to work in the BC mode, as indicated by the output lines in Fig. 5-25. In this mode, equation (5-8) is in effect. Plug in the different values for $R1$ and $R2$ ($R1$ at 10 kΩ and $R2$ at 10 kΩ and 100 kΩ) and see how the math works.

When you reverse your measurement, like when you did the AB test, you also have to turn the equation upside down:

$$V_{\text{out}} = V_{\text{in}} \times \left(\frac{R1}{R1 + R2} \right) \qquad (5\text{-}11)$$

Now add the second voltage bridge as shown in Fig. 5-26. Since $R3$ equals $R4$, the voltage at their junction is half of the input voltage. Now test the

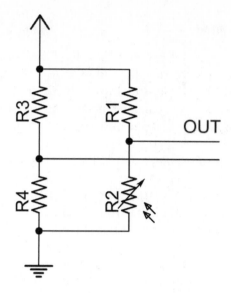

Fig. 5-26.   Two voltage bridges.

voltage between these two dividers. It is unlikely that the result is zero in the dark.

We need to add a trim potentiometer to help calibrate the circuit. Figure 5-24 shows this trim potentiometer in parallel with R3 and R4. You could simply replace R3 and R4 with the potentiometer; however, you get more sensitive calibration results if you build the circuit as shown. The top half of the trim pot is in parallel with R3 and the bottom half is in parallel with R4. Look back at equation (5-6) and see what this means for the total resistance of that voltage divider.

One question that should have crossed your mind is, how do you choose the values for R1, R3, and R4? I'm glad you asked.

Trial and error plays a part. However, if you use your multimeter to test the photoresistor, you will find that its resistance drops to about 10 kΩ in full light. I used this minimum resistance as the rating for the other resistors.

What kind of results would you get if R1 were 1 kΩ? 100 kΩ? Try these values out in equation (5-8) and see what happens.

Try This: Take your resistors and design several different series and parallel circuits. Using Ohm's Law, calculate what the resistance across the circuit should be. Now build the circuit and measure it! How close are your results to the calculations? Is it within the tolerance of the resistors?

LIGHT BULB

Let's look at one last type of resistor—the light bulb. Yes, light bulbs are resistors. They get hot, so hot that they glow white. When you heat up your electric oven, it creates light too, but mostly in the red and infrared (below red, as heat) range. See if you can find a 60-watt light bulb in your house. Clean the contacts on it and use your multimeter to find the resistance across the bulb.

In my bulb, I recorded a resistance of 19 Ω. Using Ohm's Law (equation (5-2)), and the knowledge that American house voltage is about 110 V, we calculate that this bulb should draw about 5.7 amps. Using the definition of power (equation (5-4)) we see that this translates into 627 watts.

That's not right! In fact, that's ten times what it should be. This is because resistors do not have the same resistance under all conditions. Most metals, for example, increase their resistance as they heat up. A light bulb drawing almost 6 amps is going to heat up a lot, and fast. Doing the math backwards, we calculate that the hot resistance of the bulb should be closer to 200 Ω.

Summary

This was a long chapter, but we had a lot of material to cover. First, we looked at schematics. These are the diagrams that we use to describe electronic circuits, the language of electronics.

A schematic is an abstract thing, an idea only. To make it real, we need to build the circuit using actual components. This construction can take place on circuit boards of different types, in the air, or in temporary breadboard arrangements.

There are new skills involved in circuit building, namely soldering. Though there isn't much call for soldered circuits in this book, when you buy kits or move on to more advanced projects you will need to know how to solder things together. And even here, there will be times when you may want to solder new wires onto a component so it will fit into your breadboard circuit.

Our first component was the resistor, a deceptively simple part. With it, however, we were able to build a Wheatstone Bridge that explores many of the features of electricity and resistance. We were given a chance to explore both Ohm's Law and the rules governing resistors in parallel circuits.

With your new knowledge you are better prepared to look at schematics and know what they mean, as well as to build the circuits described within them.

Quiz

1. If I draw an abstract picture of my electronic circuit, what would I call it?
2. What do these labels refer to: Gnd, R1, C5, S2?
3. What kind of board would you build a temporary circuit on? What would you call this temporary circuit? Once you had it working, what kind of board would you want to rebuild it on?
4. What is the component that opposes a circuit's current?
5. Whose formula describes the behavior of the resistor? What are some variations on this formula?
6. Say you have an LED in a 9-volt circuit that can only take 15 milli-amps of current. What size resistor would you put in series with the LED to keep it from burning out?

   What if you had two 1.1 kΩ resistors, could you make those work? How much current would the LED get in this case?
7. Draw a Wheatstone Bridge, and describe what it is good for.
8. What does electronics have to do with robots?

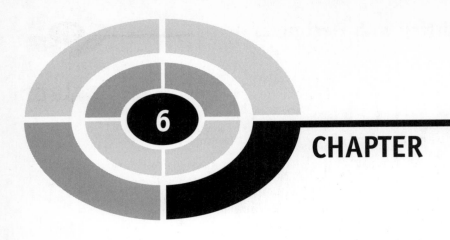

CHAPTER

6

Control

Introduction

What makes a robot interesting is its behavior, and how that behavior adapts to, or can be reprogrammed for, changing environments. The source of behavior is the control system of the robot.

When we think of robotic control, it is normal to think about computers and complicated electronics. However, the best behavioral control system is one that works automatically, without having to resort to electronics or computers.

There were control systems long before there were computers, or even electronics. This short chapter explores some of the concepts behind control systems.

Passive Control

The best control system is one that doesn't have to do anything, where the laws of physics do all of the work automatically.

For example, think about balance. You can stand a pencil on its eraser and it will stay there. Gravity is pushing straight down through the pencil and the center of gravity is over the eraser. It could stand there for a hundred years. But it is not stable.

Tip the pencil just a little bit. Perhaps you bump the desk or somebody can't resist the urge to blow on the pencil, and it falls over. Figure 6-1 shows this condition, where the center of gravity gets moved away from the eraser. When that happens, gravity's acceleration pulls that center of mass down to the table. The pencil's natural "behavior" takes over and you can rely on the force of gravity to make the pencil fall each and every time.

What if you wanted to make something that would *not* tip over like the pencil? How could you arrange the system so that gravity would make it return to its original position instead of tipping it over?

The secret goes all the way back to Chapter 2. The pencil is full of potential energy, equal to the height of its center of gravity. Given any opportunity,

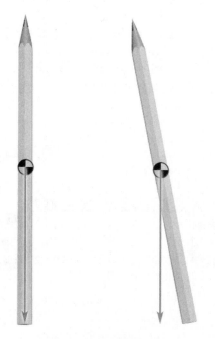

Fig. 6-1. Falling pencil.

this energy will be spent and the pencil will end up with the lowest possible energy, flat on the table.

We need to shape the system so that the center of gravity is below the pivot point instead of above it. That way, when we bump it, the center of gravity moves *up* instead of down, and gravity acts to correct the bump.

BALANCING MACHINE

Try This: Figure 6-2 shows a machine that uses gravity control to keep it stable, balanced on the tip of our previous pencil. Of course, you could also tape two pencils together in an upside-down "V" shape (like "Λ") and balance that instead. When this machine tips left or right the center of gravity rises and is accelerated back down until it is directly under the pivot point.

Open-Loop Control

The automata of the nineteenth century had complicated control systems. Inside of their mechanisms were intricate gears and metal drums covered with studs. These then bumped feelers, turned cams, and plucked levers as they turned. These preprogrammed motions were translated into dancing, writing, or music, depending on the mechanism. Similarly, the punched cards that encoded weaving patterns for looms were read and operated on by complicated control systems.

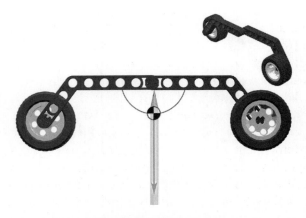

Fig. 6-2.   Better balance.

All of these actions and results occurred in a void, like the sound of a tree falling in the forest even though there is no one to hear it. The commands from the cams or the cards are sent out into the mechanism, never to be heard from again.

Did they succeed? Is the machine still working, or did something break? It doesn't matter, because in the next tick of the gears another command will be sent.

This type of control system is called *open-loop control*. Information goes in a straight line, out from the control system and into the machine. There is no information coming back, the loop is not closed. To close this loop, we need feedback.

Feedback Control

The examples in the Passive Control section actually illustrate the concept of *feedback*.

Many systems can have complex behavior without feedback. Many of the intricate automata from the nineteenth century performed complex and intricate actions based entirely on commands from their "programs," the notched wheels or studded drums that revolved in their bowels.

Many robots interact without obvious feedback, going through actions based on a prerecorded sequence—the animatronics at Disneyland and the welding robot of the factories, for example. Internally there may be hidden feedback loops that keep the mechanism on track.

There are two kinds of feedback. *Positive feedback* was illustrated by the pencil in Fig. 6-1. A small disturbance (the pencil tipping) is amplified into a larger disturbance (the pencil tipping more), continuing until the system reaches its limit (the pencil hits the table). A familiar and uncomfortable form of positive feedback is the squeal of sound you get when a microphone is too close to its speaker.

Negative feedback was illustrated by the device in Fig. 6-2. A small disturbance (tipping) creates an automatic force in the other direction, resisting the tipping. The system will *oscillate*, or wobble, back and forth until the energy put into the system is lost as heat. Feedback or, more specifically, negative feedback, is important for many if not most automatic systems.

In the eighteenth century steam engines were becoming popular, turning heat into work. These devices powered the industrial revolution and helped propel the world forward to the technological prowess we enjoy today.

In a steam engine, heat is applied to a tank of water. The water heats up and expands as steam, which creates pressure in the tank. This pressure then pushes pistons, which perform work. The question is, how do you control the steam flow to the piston to keep the machine moving at a steady speed?

CENTRIFUGAL FEEDBACK

Try This: James Watt is credited with perfecting, or at least popularizing, the *centrifugal*, or fly-ball, governor. An example of this mechanism is shown in Fig. 6-3. The block at the bottom is a standard motor from the Mindstorms kit. When you turn it on, the shaft spins and the wheel weights try to fly out away from the shaft. This, in turn, causes the lift arms to, well, lift. These are linked to a sliding link on the shaft. This sliding link can

Fig. 6-3.   Centrifugal governor.

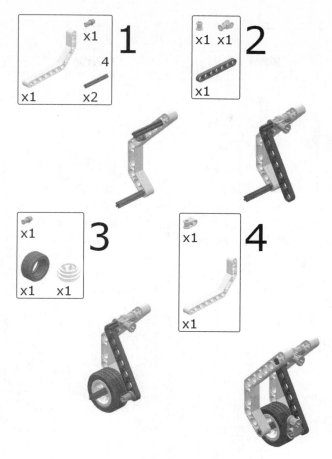

Fig. 6-4. Building the arm.

then be used to trigger or adjust the system that makes the shaft spin. The steps for building this centrifugal governor model are given in Figs. 6-4 and 6-5.

A control system that receives a feedback signal from the mechanism under control is called a *closed-loop control*. The controller sends a command to the machine and then a sensor or mechanism in the machine sends information back to the controller. The circle of communication is complete, the loop is closed.

If you build a robot car with bumpers, so that the car backs up and turns when it hits a wall, you are using feedback. Your body is constantly using feedback as it operates, to keep itself warm or cool, fed or watered. Where else do you use feedback?

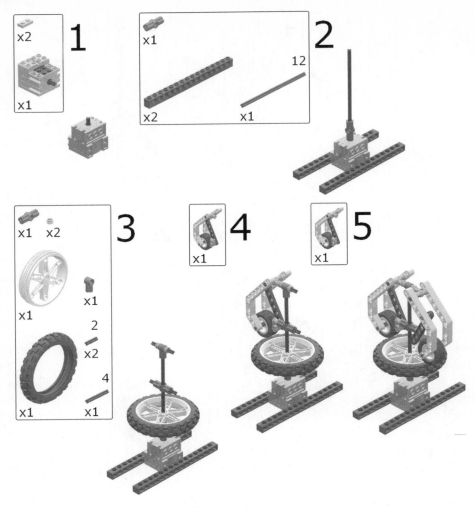

Fig. 6-5.   Building the governor.

Hysteresis

Think back to the hypothetical robot car with the bumper. Assume, for a minute, that the car goes into reverse whenever the bumper is pressed, and goes forward when it is not pressed. What do you think happens?

Your car is happily motoring forward when it hits a wall. The bumper senses this event and the car backs up—for a fraction of a second until the bumper releases. You haven't escaped the wall. Instead, the car is going to

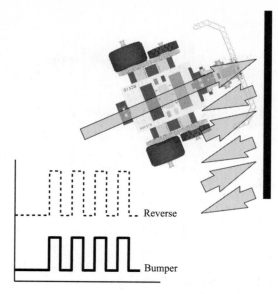

Fig. 6-6. Car bumper with immediate response.

twitch back and forth until random forces make it turn away from the wall or the batteries run down. This is illustrated in Fig. 6-6.

In the graph below the picture there are two lines that represent different events. The bottom, solid, line is the bumper signal. When it is high the bumper is pressed. The top, dashed, line is the reverse signal to the car. When it is high the car is going backward and, preferably, in a slight circle so it avoids the obstacle. Clearly, this strategy is not working very efficiently.

What you need is some kind of delay between the sensed event (the bumper signal) and its effect (the reverse signal), so that the released bumper doesn't let the car go forward again until after a sufficient delay. This delay between cause and effect is called *hysteresis*, and it is important to many control systems. An example of this is shown in Fig. 6-7.

When the bumper hits the wall the car immediately starts to back up. The bumper moves away from the wall and its signal returns to normal but, with hysteresis in the system, the car continues to back up for a predetermined time delay *t*. This provides enough of a retreat that when the car starts moving forward again it misses the wall, at least for a while.

Another example of hysteresis is in your home thermostat. These have two set points, one to turn the heater on, for example, and another to turn the heater off (Fig. 6-8). When the room temperature drops below the *Turn On* point, the heater turns on. This is an all-or-nothing proposition, there is no "medium" setting on your house heater. With the heater on, the temperature

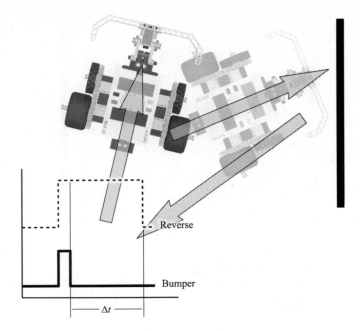

Fig. 6-7.   Car bumper with hysteresis.

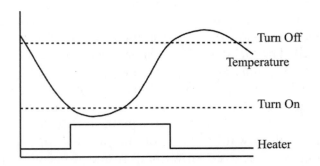

Fig. 6-8.   Temperature control.

stops its fall and begins to rise again. It quickly passes the *Turn On* point, but the heater remains on until it passes the *Turn Off* point.

MECHANICAL SWITCH

Try This: Our third, and final, example of hysteresis is a mechanical switch. It is similar to the falling pencil example earlier in the chapter, but has two states, "up" and "down," instead of just one, "fallen." See Fig. 6-9.

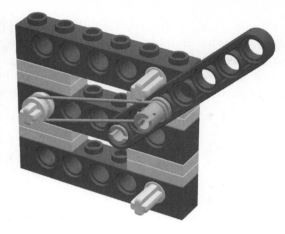

Fig. 6-9. Mechanical switch.

The rubber band holds the lever against the post in an *energy well*. It takes energy to move the lever, since the arc of motion pulls against the rubber band. Putting enough energy into the lever causes the lever to escape its well. However, the pull of the rubber band snaps it down into the other energy well. There it sits until disturbed again.

This arrangement of force is useful for holding something in place, without having to use motors to keep it there. The only time you need to use force is to switch the position, or state, of the switch. Try to extend this design into a mechanical gripping hand that stays closed under its own power.

Summary

In this chapter we continued to lay down a foundation of concepts for our later work. We started with passive control, a form of control where you don't have to do anything to manage it.

From there we looked at some active control concepts, starting with open-loop control where the controller doesn't have any confirmation, or feedback, as to what the system is doing. A more powerful form of control is closed-loop control, where feedback is used to adjust the system.

Feedback doesn't imply fancy sensors, electronics, or computers. Very simple systems use feedback to good effect, and these simple principles can be used in more complicated systems as well. Whenever you can make it work, simple mechanisms are better than complex ones—less likely to break, easier to understand, and more likely to work throughout the life of the system.

Every time you think up a complicated solution to a robotic problem, take a moment and see if you can find a simpler way of doing it. How would someone from the steam-powered age handle it?

Finally, we looked at how delays in a controller can be important. Fast response is not always useful, so we added hysteresis, or delays, to our bag of tricks.

Quiz

1. According to the author, what makes a robot interesting?
2. What kind of control systems did the nineteenth-century automata use?
3. Fancier robots need fancier control systems. What is another, more capable, type of control?

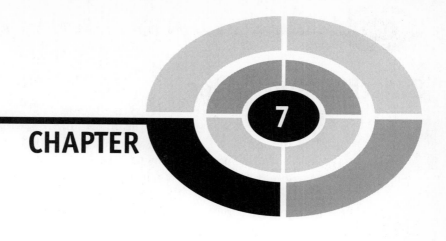

Sequencing and Programs

Introduction

While basic feedback control holds a valuable place in the spectrum of control systems, most robotic systems need more complex control solutions.

Many robot tasks require the machine to perform a preprogrammed *sequence* of behaviors, one thing after another and so on until the pattern is complete. A sequence of behavior can be defined by a *program*, which is a list of instructions telling the machine what to do.

We shall turn to the computer, the ultimate in reprogrammable controllers, to develop these control programs. There are also simple, mechanical ways to create a sequential program, so we look at these first.

Switches and Cycles

One of the easiest ways to sense the environment is with a *switch*. We already implied the use of an electrical switch in Chapter 6, with the car's bumper. We also saw a nonelectrical switch in Fig. 6-9.

A switch, in electrical terms, is a mechanical device that can make or break an electrical connection. There are many different types of switch, but the one we look at here is mechanically actuated, that is, you change it by pushing on a lever or button.

Figure 7-1 shows switches that can sense the position of their respective machines. The top machine rotates. As it rotates, a lever spins around and pushes on a lever on either the left-hand switch or the right-hand switch.

The rotating machine has a flaw in the way the switches are placed. If the machine *overshoots*, or goes too far, it may bend the lever or break the switch. The machine at the bottom of Fig. 7-1 has a sliding part. If this machine overshoots, it simply passes underneath the switch with no risk of breaking it. Note how the levers on the switches are pointing in opposite directions. Why are they doing that?

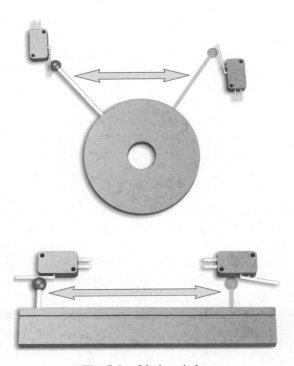

Fig. 7-1. Limit switches.

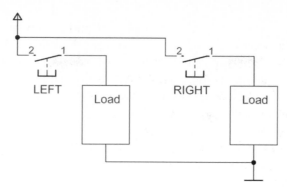

Fig. 7-2. Switch schematic.

A schematic for both machines in Fig. 7-1 is given in Fig. 7-2. There are a few new symbols in this schematic. Note, for example, the alternative ground symbol in the bottom-right corner. You can't let subtle changes in symbology fluster you, since there are many different ways of drawing the same thing. If it looks like a duck and quacks like a duck, but smells just a bit fishy, it's still probably a duck.

The switches are along the top. These are *normally open* switches, meaning that when no other forces are acting on them they create an open circuit. No electricity can flow. Once you push the button (or lever, or what have you), the circuit closes and electricity flows through the switch. A *normally closed* switch is the opposite. Electricity can flow through it until the button is pushed.

Switches that are used to mark the extremes edges of a motion are known as *limit switches*, since they guard the limits, or ends, of the motion. In this schematic, the left-hand switch closes when the machine is in the *LEFT* position, and the right-hand switch is for the *RIGHT* position. The little numbers on the switches refer to the terminal numbers, which may be printed on the switch itself. They have no other meaning.

The big blank boxes labeled *load* can be anything. A load is something that puts a load on the circuit, using some of its power. It could be a resistor, a motor, or some other device. All we can tell from this circuit is that when the *LEFT* switch closes it powers the left-hand load, and when the *RIGHT* switch closes it powers the right-hand load.

If the left-hand load is a motor that makes the machine move to the right, and the right-hand load is a motor that makes the machine move to the left, what would happen? Not much. But assume that a motor stays on until another motor starts. What then? This requires a more complex circuit than the one shown, but we can gloss over these issue for these examples.

The machine would *cycle*. A cycle is a round of events that happens repeatedly in the same order, a loop of action. The cycle described above would have the machine shift to the right until it hit the switch, shift back to the left switches, and then repeat until the power was turned off.

Using nothing more than mechanical switches, you can create long chains of actions. For example, imagine the robot arm shown in Fig. 7-3. Assume that it has four actions: the arm can retract (get shorter), extend (get longer), grab, and drop. Each motion has a limit switch labeled *RETRACTED*, *EXTENDED*, *CLOSED*, and *OPEN*. We assume that it begins in the retracted and open position. What will the arm do if it is hooked up to the control system in Fig. 7-4?

Note that these systems also assume that the machine starts in a position that forces a switch closed. While we can't normally rely on the initial position of a machine, we can ignore the complexities of the real world for the sake of illustration.

So, in this perfect universe, the arm should extend to its outer limit, close the gripper, retract, and then open the gripper again. This four-step, or four-*state*, cycle will repeat forever. A state, by the way, is a combination of attributes—in this case, the position of the switches and the power applied to the loads. Only the significant states, where there is a change, count.

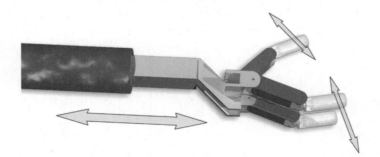

Fig. 7-3. Grabbing arm.

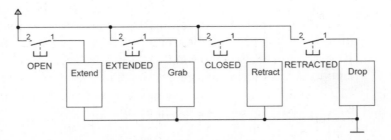

Fig. 7-4. Grabbing schematic.

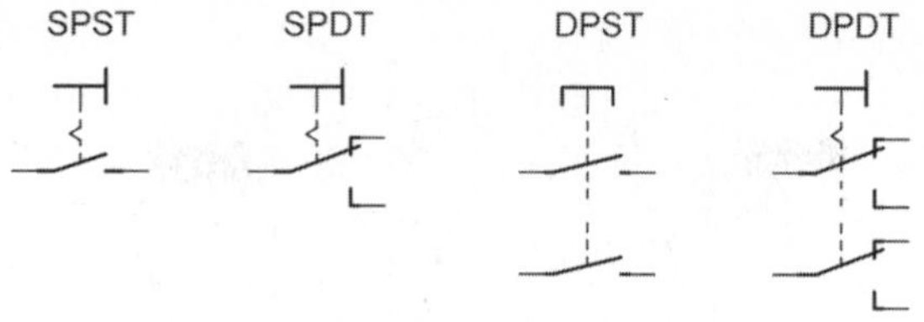

Fig. 7-5. Switches.

To wrap up this section, let's look at some more switches. A switch is defined by how many circuits, or *poles*, it controls and how many states or *throws* it has. The switches we have looked at so far are *single-pole* and *single-throw*, or *SPST*. Figure 7-5 shows the four common switch configurations.

The SPST switch you should already recognize. Note that this is shown as a *normally open* (N.O.) switch. SPST switches also come in *normally closed* (N.C.), which starts closed and opens the circuit when it is switched.

The next switch is *double-throw*, making it an *SPDT* switch. The pole is the switching part and there is still just one of these. However, the switch makes a different connection in each of its two positions. One of the connections may be N.O. and then the other is N.C. In practice there may not be a distinction unless the switch is spring-loaded to return to a standard "normal" position.

If you put two *SPST* switches into one package so they share a switching mechanism, you get a *DPST* switch. Likewise, a pair of *SPDT* switches work together to make a *DPDT* switch.

Switches may toggle, latching onto their new state after they are adjusted, or they may be spring-loaded so they return to a default position. Switches that return to their default are called *momentary contact* switches. They perform their switching action while being pushed on but return to a default state when the influence passes.

The switching mechanism may be a button or lever, converting mechanical motion into an electrical connection. However, there are other ways of operating a switch. A vial of mercury with two wires poking into it, for example, makes a simple tilt-powered switch. Air pressure, electricity, and vibration can all be used to control switches.

Cam Control

Control information can be encoded on a *cam*. A cam is a wheel with one or more projections. In the simplest form, a cam is a round wheel mounted off-center, like the *eccentric cam* shown in Fig. 7-6. Eccentric doesn't mean

Fig. 7-6.   Eccentric cam.

that it lives alone in a big house collecting cats. It literally means that it is "off-center."

The cam provides one way to convert rotary motion (the turning of the cam) into linear motion. Look at the cam from a fixed point of view. As the cam rotates its edge gets closer and closer to us, until it reaches its closest point and then the edge retreats.

This moving edge can be used to trigger a switch, push a lever, or slide a bar. When the cam pushes on a lever or rod, the pushed piece is called a *cam follower*. Followers may have rollers or some other friction-reducing scheme on their end to keep them from rubbing on the cam and slowing things down.

Cams don't have to be round, but can have any shape desired. Cams may also consist of pins or pegs stuck into a wheel, so long as the cam follower doesn't get stuck on them. Figure 7-7 shows a more intricate cam, a cam follower, and the motion of the follower traced out on a graph.

Multiple cams can be stacked on the same shaft, each one controlling some aspect of a machine. The cams don't have to be individual disks, but can be a solid drum or bar of material that has been shaped to act like a series of cams. Music boxes use this type of mechanism, a rotating drum with pins on it that pluck the musical tines. The only limit to a cam's shape is your imagination.

Changing the programming on a cam-driven machine requires that you take it apart and replace pieces inside of it. This may not be the most efficient way to change the machine's action. More advanced cams let you replace their bumps without taking things entirely apart, but this is still a clumsy way to program a robot.

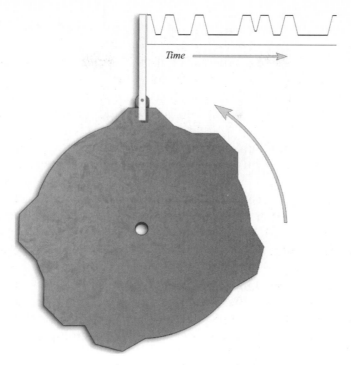

Fig. 7-7. Cam follower.

CARDBOARD CAM

Try This: Draw a cam shape onto a piece of heavy paper or cardboard. Cut it out! Punch a hole in the middle and stick a pencil through it. Now tape a piece of paper to the wall. If you hold the point of the cam's axle pencil against the paper and then roll the cam, you should be able to trace out the control signal on the paper.

Did you get what you expect? Of course, if you have two bumps close together, they will be merged together by the large flat surface of the floor. The floor doesn't make a very precise cam follower. Try drawing a control signal and then designing a cam that will reproduce it.

Card Control

While cam control works, it can be hard to reprogram. What we need is some way to feed instructions to a machine from the outside. Then we can feed it different instructions at different times, to make it perform different actions.

First, let's make a simplifying assumption. Cams have a special property that is so obvious that you probably didn't even notice it. Their bumps can be of any height or shape. If you traced the graph on the eccentric cam from Fig. 7-6, for example, you would get a smooth sinusoidal, or wave-like, shape. The graph for Fig. 7-7 had a more structured look, but the fact remains that we could make our graph take almost any shape we want. This makes the output from cams *analog*, meaning the cam creates a smooth, continuous signal.

If we assume, instead, that we are triggering a switch with the control cam, we only care if the switch is *on* or *off*. This is a *digital* signal, meaning it has individual discrete states that you can count on your fingers (digits).

Assuming we want a digital control signal, there are a number of options open to us. One of the oldest digital control systems is the *punched card*. This is a card, made of thin wood or thick paper, with holes punched in it. An example is shown in Fig. 7-8. The small rectangles in the figure are holes in the card. When the card passes through a reader, small fingers make contact through the holes. The reader is, in fact, a multipole card-actuated switch.

There are versions of the card reader that don't use mechanical switches. Light passing through the holes to a sensor is faster and doesn't wear down the card. Those multiple-choice tests where you fill in the bubble with a pencil are a descendent of the punch card.

Joseph Marie Jacquard made good use of punched cards to control patterns in a silk loom in the early nineteenth century. His cards were linked into a long belt so the pattern could cycle endlessly, or until something jammed.

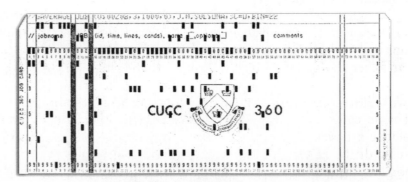

Fig. 7-8. Punched card.

Herman Hollerith used loose cards to tally statistics during the 1890 U.S. census. He then went on to patent his card-based tabulating machine and founded the Tabulating Machine Company which later became International Business Machines or, as we know it now, IBM.

Related to the punched card is punched tape. This is like a long, skinny card. Holes punched into the tape code numbers, and the numbers in turn encode instructions. Punched tape was used until fairly recently to control automatic machine tools. The instructions on the tape described the shapes to be cut by the machine. Both punched cards and punched tape have fallen into disuse lately, being replaced by the magnetic and optical storage used in electronic computers.

MECHANICAL CARD READER

Try This: You can create a mechanical "card reader" that, instead of sensing holes in a card, reacts to bumps on a plate, not unlike a flattened cam-and-follower arrangement.

Figure 7-9 shows the construction steps for this reader. Note that in step 3, the two liftarms are not locked together. Make sure that everything is assembled loosely enough so that the liftarms move smoothly and independently. If gravity isn't strong enough for your liking, you can loop rubber bands across the obvious attachment points.

Figure 7-10 shows a bump-based card. If you take this card and slide it through the reader, the bumps make the liftarms lift. If you have smooth plates in your LEGO collection, you can improve the action a lot by smoothing over the bumps in the reader's path. If you wanted to make this reader control something, you could put switches where the top ends of the liftarms would trigger them. Note that this mechanism could be adapted to make a cam follower.

Programming Concepts

What we really want is a way to control a machine without needing to create special cams, cards, or tapes. We want to be able to plug in a cable and send the machine a bunch of instructions, changing its programming like magic. In the case of the LEGO RCX, you can set the controller brick next to the data port and the transfer is even more magical, done using low-frequency light.

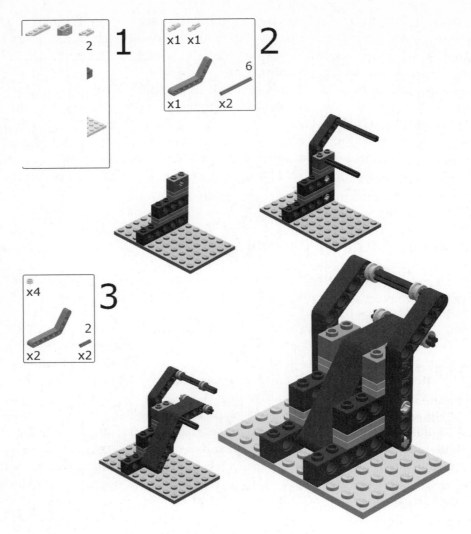

Fig. 7-9. LEGO "card" reader.

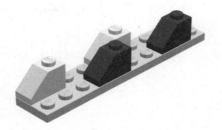

Fig. 7-10. LEGO "card."

When we use a cam to control a mechanical link, it's obvious what is happening. Using that same cam to flip a switch is still pretty clear. The switch will turn on a motor, turn off a light, start a timer, or perform some other directly linked action. And since punched cards and tape are like paper-driven switches, its not to hard to imagine how they can control a machine as well.

But not all controllers do their job directly, switch-to-actuator. Note that an *actuator* is a mechanism that a controlling agent uses to operate on its environment—something that acts. Sometimes the holes in the card or tape encode instructions to a computer, and the computer uses these instructions or *programs* to control the robot indirectly.

The first step is to convert the holes in a card into numbers. Then, we need to make the numbers mean something to the computer. Finally we can use the numbers and the built-in capabilities of the computer to control our equipment. The design of a computer is closely tied to how we represent our numbers, so describing this system is something of a chicken-and-egg problem. We start with the numbers.

COMPUTER NUMBERS

Computers operate in the world of "on" and "off," also known as *binary*. As a simple example, look at the schematic in Fig. 7-11. This is a battery (the sandwich-shaped symbol on the left), a switch, and a light-emitting diode (LED)/resistor pair. When the switch is open, its default position, the LED is off. If you push the button the LED turns on. This display is telling us the state of our switch and represents a single binary unit, or *bit*.

What if you have four bits, as shown in Fig. 7-12? Each LED, or bit, could represent a single state in the system, giving four states. If we allow more than one LED to be on at a time, we have sixteen unique states defined by the four lights. The sixteen patterns of four bits are given in the left column of Table 7-1. Note the other columns show different interpretations of these patterns.

The binary column exactly mimics the lights but with 0 for a dark bulb and a 1 for a lit one. For easy reference, each binary pattern has a convenient label, 0 through F. "F?" you might ask. "F" provides a handy way to say "15" with a single symbol, and is used in hexadecimal numbers. *Hexadecimal* is base 16 ("hexe" for six plus "decimal" for ten), because there are 16 patterns.

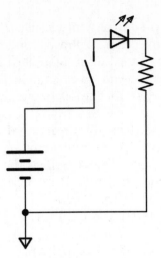

Fig. 7-11.   One bit.

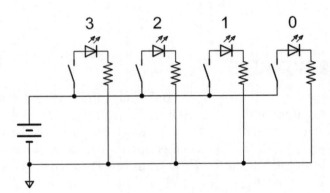

Fig. 7-12.   Four bits.

In early computers, scientists resisted moving beyond our familiar digits 0 through 9. They used only three bits to get eight patterns. These days we use four bits, and this four-bit pattern is called a *nibble*. Two nibbles make a *byte*, two bytes a *word*, and two words a double-word or *dword* which contains 32 bits.

Fortunately for you, the discussion of number systems and the intricate details of binary are beyond the scope of this book. The following box does give a brief introduction for the mathematically inclined.

The binary system

The numbers we use in daily life consist of ten digits. The digit 1 represents a single thing, such as a pebble. 2 is a pebble next to it. Add another pebble and you can represent their quantity with the symbol 3. The digits are just symbols that stand in for a particular quantity of pebbles. 0 is special since it represents no pebbles.

Nine pebbles are the most we can represent with a single digit. What happens when we add another pebble? We get the symbol sequence 10. Each digit's value is affected by its position in the sequence. The right-most digit is in the "ones" place and is worth its face value. The next digit to the left is in the "tens" place, where a 1 stands for ten pebbles, 2 is twenty, and so forth.

This system is known as the decimal system, or base-10. Each place value is ten times the place value to its right. To make the obvious complicated, the quantity represented by the sequence of digits is calculated by:

$$q_{10} = \sum_{p=0}^{n-1} d_p \times 10^p$$

where:

q_{10} is the quantity represented by the sequence of base-10 digits;
n is the number of digits in the sequence;
p is the place number of a digit, from zero at the right to $n-1$ at the left;
d is the quantity represented by the digit at position p.

Binary coding is exactly the same but with only two digits, 0 and 1:

$$q_2 = \sum_{p=0}^{n-1} d_p \times 2^p$$

So the binary sequence:

$$1101011$$

represents the quantity:

$$\left(1 \times 2^6\right) + \left(1 \times 2^5\right) + \left(0 \times 2^4\right) + \left(1 \times 2^3\right)$$
$$+ \left(0 \times 2^2\right) + \left(1 \times 2^1\right) + \left(1 \times 2^0\right)$$
$$= 64 + 32 + 8 + 2 + 1$$
$$= 107$$

From Edwin Wise, *Hands-On AI with Java* (McGraw-Hill, 2004)

Table 7-1   Four-bit patterns and labeling

Bit (Light) Pattern	Binary	Label
● ● ● ●	0000	0
● ● ● ✸	0001	1
● ● ✸ ●	0010	2
● ● ✸ ✸	0011	3
● ✸ ● ●	0100	4
● ✸ ● ✸	0101	5
● ✸ ✸ ●	0110	6
● ✸ ✸ ✸	0111	7
✸ ● ● ●	1000	8
✸ ● ● ✸	1001	9
✸ ● ✸ ●	1010	A
✸ ● ✸ ✸	1011	B
✸ ✸ ● ●	1100	C
✸ ✸ ● ✸	1101	D
✸ ✸ ✸ ●	1110	E
✸ ✸ ✸ ✸	1111	F

For now you can take it as a matter of faith that patterns of on and off can be read as numbers. Each number stands for a specific pattern of lights.

Of course, in electronics we don't have to use LEDs and mechanical switches to represent bits. We can use more subtle components—transistors for switches, pulses of electrons against a phosphor-covered tube for indicators (your TV or monitor), magnetic fields on spinning sheets of magnetized plastic (floppy disks), pits burned into a metal film laminated onto a plastic disk (CDs), and so forth. Or, going back a ways, holes in cards.

COMPUTER INSTRUCTIONS

Patterns of on and off can represent numbers. Likewise, numbers can represent other things.

For example, the letter A can be represented by the number 0, B by 1, C by 2, and so on. In fact, A is normally represented by the number 65 and B by 66. Lower-case a is 97. This particular letter coding is part of the American Standard Code for Information Interchange, or *ASCII*, code. In ASCII, my name "Edwin" is the sequence of numbers 69, 100, 119, 105, 110.

It's not much of a reach, then, to imagine that we can use numbers to represent commands to the computer. 0 could be, for example, *halt*. 1 might be *load a value*, and 2 *store a value*.

It would be the subject of a whole book, if not more, to trace the progression from switches and lights to full computers. However, any digital computer you have ever used operates using a small handful of very simple operations, performed by electronic circuits, on values of one and zero.

The holes punched into tape or cards were read by little switches. These switches created voltages inside the early computers, and these voltages were stored in circuits. Sometimes the circuits would cause the voltages to be converted into magnetic patterns on magnetic disks or tapes, other times they might trigger machinery to punch holes in some tape.

The patterns of voltages, the patterns of on and off, flow through the computer's circuits to create different patterns of on and off. This all happens very quickly, since electrons act very quickly, and can seem like magic. But inside the computer there are only many tiny electronic switches, switching on and off.

LEGO RCX

Leaping ahead several decades of research and development, you will find that numbers can represent complex instructions to a robot, for example. Looking at Fig. 7-13 you can see four "blocks" connected together. This is one way to represent a program that is easier on the eyes than a stream of numbers, binary, hex, or otherwise. Underneath it all, however, it is still a set of binary patterns.

The top block represents the starting point, which for this simple example is labeled "Untitled." The next block represents a command to the robot to

Fig. 7-13. Action blocks.

move forward for one second. Then you have the command to turn left for a second, and finally another second of forward motion.

If you have used the LEGO Mindstorms at all, you will recognize this as a primitive program. You can even build the Roverbot in the Mindstorm instructions and run this program on it.

Thinking back to the beginnings of this chapter, this program could be created by two cams on a slowly moving shaft.

The first cam would be the "forward" cam, and would have a bump that would press a switch that makes the robot move forward by supplying power to both of its motors. The size of these bumps, for there would be two of them, would be large enough so the switch was pressed for exactly one second.

The second cam would be the "turn left" cam. Its switch could make the robot turn left by supplying power to only the right-hand motor.

How would this two-cam system look? How would you wire the switches to power the motors to make it match the behavior of this program?

Summary

We started by looking at how limit switches can be used to affect the behavior of a mechanical system. In the process, we learned about different kinds of switches. By moving the switches away from the machine and then controlling them using a set of rotating cams, we saw how we might control behavior in more complex ways.

Of course, rebuilding parts of your machine to change the programming is cumbersome, so we looked at different ways to represent our sequences of on and off. This led us to punched cards and tape.

From there we looked at how patterns of on and off can represent numbers, and how these numbers can be used to represent other things, such as instructions to a computer. And what are instructions to a computer if not a program? So programs are numbers, which are simply patterns of zero and one. And what are "zero" and "one" if not the open and closed position of switches? And we are back to the beginning again.

Quiz

1. List as many types of switches as you can.
2. What is a common mechanism used to control automata?
3. How were early computers, and even earlier looms, programmed?
4. You walk into your bathroom and see that there are eight lights above the mirror. From left to right, the light bulbs are On, Off, On, On, Off, Off, On, On. Write this down in binary and decode it. What number does it represent?
5. If you were going to invent a simple computer language, what types of commands would you use?

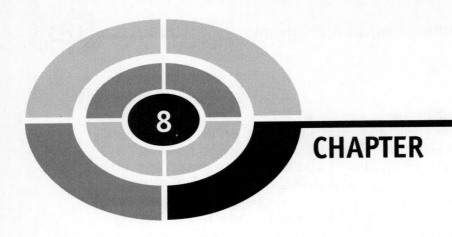

CHAPTER 8

Joints

Introduction

Chapter 3 focused on machines that manipulate force. These machines implied motion—the pivot of the lever, the turning of the gear, even the sliding of the block on a ramp. But these motions were secondary to the main focus, which was the study of the forces involved. This chapter, and the next, focus on the motion itself.

While a loose object moves when you apply force to it, robotic machines need more controlled motion. The pieces all need to be connected, and their motions constrained into useful patterns.

At the point where two rigid parts are attached, where they move relative to each other, there is a moving *joint*. This chapter is concerned with these moving joints, such as the hinge and the pivot. The basic types of joint allow bending motion, sliding, rotation, and even some complex movements.

Rotation and Bending

ROTATION

The simplest type of joint provides a simple rotation around one axis (Fig. 8-1). If you recall from Chapter 2, an axis is a direction, or vector, in space. In this context, it is the pivot upon which we turn. An *axle*. Wheels spin around axles and provide a familiar example of rotation.

The axle is so common that we don't even notice it. A rough pivot can be made by hammering a nail through two boards. Better versions provide one or two smooth holes with a smooth rod to turn in them.

Rotation occurs around a pivot, axle, pin, or shaft. These are all cylinders that pass through holes in the parts. The pivot may be free-floating or it could be rigidly fastened to one of the pieces.

Figure 8-2 shows a number of different pivots. These are general purpose, and wouldn't work for a high-speed or heavy machine. A bolt is shown in the top-left corner of the figure. Bolts are not good to use as axles in most places. They are designed to hold things together in tension, not to provide a pivot point. However, smooth shaft bolts like the one pictured can be used for light-duty projects. The washers on both ends keep the head and nut from catching on the pivoting material.

Under the bolt is a pin with a *shaft collar* on it. The collar has a screw set into its side (a *set screw*) that passes through the collar and presses into the pin. This keeps the collar from sliding off, so your machine stays in one piece.

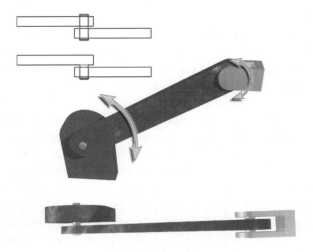

Fig. 8-1. Pivot joint.

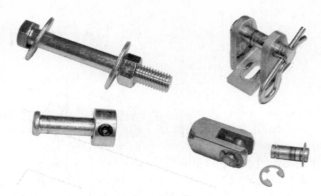

Fig. 8-2. Pivot examples.

Any cylinder of metal can be used as a pivot, though for heavy jobs you need to have good hard metal. If you need a large pivot, you can even use hollow tubing.

The right side of the figure shows a clevis. A *clevis* is a U-shaped piece of metal with holes in its ends. A pin passes through these holds to attach some bar to the clevis. The clevis to the right of the bolt fastened a pneumatic cylinder to a base plate. The pin passes through the mounting clevis, a hole in the base of the cylinder, and back out of the clevis. A wire threads through a hole in the end of the pin to keep it in place.

The pin could be held in place with spring clips, as shown in the bottom-right corner of Fig. 8-2. The spring snaps into a narrow groove around the ends of the pin.

BEARINGS AND BUSHINGS

If your machine is heavy, the friction between the pivot and the part can get large. In that case, you need to take extra steps to reduce the friction of the moving parts.

The simplest tool to reduce friction is the *bushing*. The bushing illustrated in the left-hand side of Fig. 8-3 is sintered bronze that has been filled with oil. Sintering is a way to make a porous solid, like a fine sponge. The technique of sintering is used in ceramic and metals and involves heating and pressing, or gluing, a bunch of particles together into a porous block.

The bronze bushing is constantly leaking its oil into the joint, keeping it running smoothly. Bushings can also be made from hard, smooth plastics like nylon or PTFE compounds. These plastics are both tough and naturally smooth.

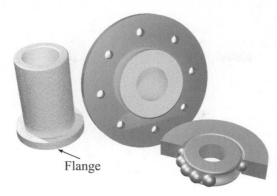

Flange

Fig. 8-3. Bushing and bearing.

To use a bushing, you drill a hole in your part that is just slightly smaller than the outside of the bushing. The bushing is pressed into the part and friction, ironically enough, holds it in place. A rotating shaft runs through the middle of the bushing.

Bushings are good for slowly rotating joints, but for high-speed operation you need something even better. *Ball bearings* are that something.

Ball bearings are hard metal spheres locked into a frame. The balls roll instead of rub. Because we aren't rubbing two parts together there is almost no friction. A ball bearing normally consists of the balls and two tracks that lock together around them, as shown in the cutaway drawing on the right side of Fig. 8-3. Like bushings, the ball-bearing assembly is pressed into a tight-fitting hole in the machine and a shaft passes through its center hole.

You can also buy bearings and bushings that have been premounted into a plate or bracket, sometimes called a *pillow block*. This block can then be bolted to your machine so you don't have to worry about the difficult process of pressing the bearing into place.

You can find rotating joints everywhere you look, from doorknobs to the wheels of your car or bicycle. Even the propeller on a beanie hat uses a simple rotating joint.

BENDING

Bending motion is stereotyped by the action of the hinge (Fig. 8-4). Your elbow bends, and so does your knee. Everything that you build with a hinge has that same bending motion. If you look at the hinge from its edge, you can see that it is actually rotating around its hinge pin. Bending is a specialized form of rotation. If you have two heavy parts that need to bend relative to

Fig. 8-4. Hinge.

each other, you can build a hinge using the bearings or bushings described in the previous section.

Sliding

While rotating and bending motions can be used for most of your mechanical needs, sometimes parts need to slide along a straight line. Some of the same tools you used for rotation can be used for sliding (Fig. 8-5). For example, the bushing works just as well for sliding motion as it does for rotation.

You can unroll a rotating ball bearing to make a sliding track. You probably have these tracks on the drawers in your kitchen or desk. The cutaway illustration in Fig. 8-5 is just that, an illustration. Actual linear bearings need a way to lock the two tracks together, so the system doesn't just fall apart.

The wheels on your car or bike provide a form of sliding motion, if you consider the movement of the car against the road instead of the motion of the wheels relative to the car. Wheels on a track are similar to a ball-bearing raceway. Roller coasters and trains use wheels that lock onto a fixed track.

Complex Motion

BALL AND SOCKET

While sliding and rotating joints cover many motions, there is one more joint to look at, the *ball and socket*. Your hip and shoulder operate on the ball and socket principle, giving your arms and legs a wide range of motion.

Fig. 8-5. Sliding.

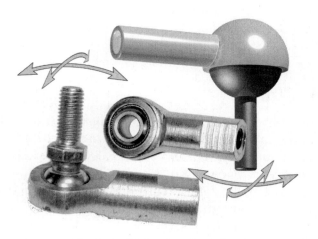

Fig. 8-6. Ball and socket joints.

Rotating and sliding mechanisms provide motion along one axis (unless you allow your slider to rotate around its shaft). They are said to have one *degree of freedom*; they are free to move along one dimension. The ball and socket has two *degrees of freedom*, since it can pivot up and down as well as back and forth.

Figure 8-6 shows the ball and socket in three different forms. The illustration in the top-right corner is a type of joint found in radio-control airplanes. The ball is attached to a threaded stud that attaches to a control horn on a servo. The socket is attached to a push-rod and runs through the airplane to push and pull the control surfaces. The depth of the socket is just over half the diameter of the ball, so the opening in the socket is smaller than the diameter of the ball. Because of this, it snaps firmly into place around the ball.

The bottom-left version is essentially the same. A brass ball is inserted into a ring-shaped socket that is curved to match the ball. The opening at both ends of the "socket ring" are smaller than the ball, and the middle opens up to match the curve of the balls. The ring snaps onto the ball.

The ball may or may not have an attachment point. It may have a hole through it, where a shaft can slide or turn. Note that in this case you could rotate the shaft in the hole, as well as rotate the ball back and forth in its ring, giving three axes of rotation.

The problem with these ball and socket joints is that they have a lot of friction; the fit needs to be tight to keep the ball from falling out. There is a lot of contact between the ball and its socket, and this creates a lot of friction.

A ball and socket give two degrees of rotation, so could we put together two single-axis joints in a way to simulate this? Yes. The result is a universal joint, or *U-joint*.

UNIVERSAL JOINT

A universal joint is two rotating joints in a single package. The two shafts are attached at a single cross-shaped *spider*. Two hinged ends attach to this, with each end providing one axis of rotation around the spider (Fig. 8-7).

The U-joint is used for power transmission. The two ends have long shafts attached to them and these rotate. Turn one shaft and this power is transmitted through the U-joint into the other shaft. When you bend the shafts relative to each other, the U-joint transmits the power around the curve. Vehicles of all kinds, from some motorcycles to cars and tractors, use some variation of the U-joint to connect their engines to their wheels.

Fig. 8-7.   Universal joint.

The standard U-joint, also called a *Cardan joint* after the Italian mathematician Geronimo Cardano, does have some drawbacks. At higher speeds of rotation it vibrates. If the two shafts are in a direct line, there is no problem. If there is an angle between them, and there will be because that's the whole point of this joint, the geometry of the joint describes a complex arc through space. This causes unwanted vibration at high speeds.

A variation on the U-joint, called the constant velocity or *CV joint*, eliminates this source of vibration. CV joints are found where the rotation needs to be fast, and the simpler U-joints where rotation is slow.

ROBOT WRIST

Try This: The machine in Fig. 8-8 consists of two rectangular frames with three rotating joints. It's a simple machine that you can build in just a few minutes.

Figure 8-9 shows the first step. The gear inside the frame is a 24-tooth crown gear. This meshes with other gears at a right angle. This gear drives the wrist plate, which is the larger gear on the outside of the frame. In theory, more parts, perhaps making up a gripping hand, could be mounted on the wrist plate. We talk about angle-turning gears like the crown gear and bevel gears in the next chapter.

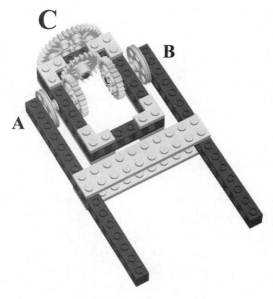

Fig. 8-8.   Wrist joint.

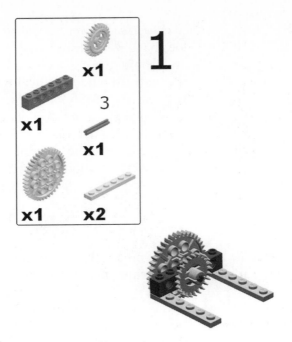

Fig. 8-9. Wrist step 1.

The other two gears, added in Fig. 8-10, also have 24 teeth. These mesh with the crown gear and, through the crown gear, with each other. Since they all have the same number of teeth there is no change in force or motion. What happens if you use larger or smaller gears here?

The last two steps, in Fig. 8-11, add small pulleys and an outer frame. The pulleys are where the power, in the form of your fingers, is applied to the machine. The axles through these pulleys and their matching gears also provide a rotating connection between the inner and outer frames.

We have two sources of power that control two different motions. One motion is the rotation of the wrist and the other is the tilt of the inner wrist frame with the outer arm. The obvious way to manage this motion might have been to have one power source control the tilt and another to control the rotation.

That's not what this machine does. You have probably noticed by now that both pulleys are involved in both motions. If you turn each pulley in opposite and equal rotations, the wrist tilts but does not turn. Note that when one pulley is moving clockwise and the other is moving counterclockwise, their top and bottom edges are moving in the same direction. Clockwise and counterclockwise is determined as you face the outside face of the pulley.

Fig. 8-10. Wrist step 2.

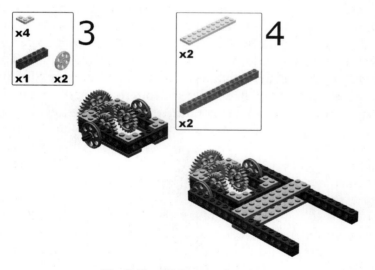

Fig. 8-11. Wrist steps 3 and 4.

If you give both pulleys the same rotation, so their top edges are moving in different directions, the wrist turns but does not tilt. Moving one pulley and holding the other still makes the wrist both turn and tilt. Different rates and directions of rotation in the two pulleys change the ratio of the turning and tilting motions.

Now that you have had some experience with this type of three-gear system, the differential gear in the next chapter may not seem so strange.

OTHERS

Try This: You should always feel free to be creative when you build your machines. Sometimes you can use things in unexpected ways. For example, a spring can be used as a multidirectional joint that has the added benefit of always snapping back to center (Fig. 8-12).

See what kinds of joints you can build from ordinary things around the house. Keep your eyes open for creative ways to use existing materials in new and interesting ways.

Summary

This chapter has been about rotation and friction. This was obvious in the first section since we talked specifically about rotation. Then, with bearings and bushings, we talked about how to make the rotation more efficient by getting rid of as much friction as possible.

Bending, we then discovered, has its roots in rotation. Sliding, at first glance, appears to be completely different from rotation. And in many ways it *is* different. However, we can use rotating joints to create low-friction sliding.

Fig. 8-12.   Spring joint.

The ball and socket joint is more like a sliding joint than a rotating one, and provides several degrees of motion. The U-joint also provides several degrees of motion, but is made up from two rotating joints. We ended the chapter by using rotating joints to create an interesting robotic wrist.

Quiz

1. What kind of joint is used by the wheel on your car? How do you reduce the friction in this type of joint?
2. How can friction be reduced in the track that guides the drawers in your kitchen? How is this joint and its friction fighters the same as, or different from, the previous one?
3. You want to power a wheel, but it is not in a line with your motor. What do you do?

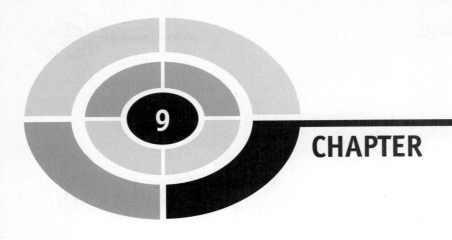

CHAPTER 9

Power Transmission

Introduction

Thinking about power transmission brings images of large metal towers carrying high-tension lines throbbing with the tamed lightning that we use to run our cities. This chapter, however, is concerned with the transmission of mechanical power.

The easiest way to move power from a motor to its load or other device, in this example a wheel, is to hook them together directly. Either attach the wheel to the motor itself or connect them with a shaft extension, that is, a long bar (Fig. 9-1). With the addition of a pair of U-joints, you can bend the shaft around corners (Fig. 9-2).

While direct drive is an important part of a power transmission system, it is rarely sufficient all by itself. This chapter combines ideas from Chapters 3 and 8 and comes up with a variety of ways to transmit force through your machine, modifying it along the way to suit the task at hand.

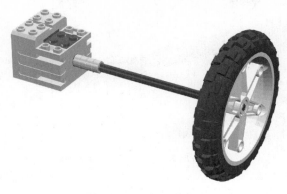

Fig. 9-1.   Direct drive.

Fig. 9-2.   Direct drive with U-joint.

Chains, Belts, and Cables

We introduced chains and sprockets at the end of Chapter 3. Figure 9-3 shows a complete chain and two sprockets.

The sprockets are gears with teeth designed to mesh with a chain instead of another gear. The chain is a series of links held together with pins. There may be rollers on the pins, acting as bearings, to reduce friction. The links in a chain come in two different forms. One form is like the links shown in Fig. 9-4, with two alternating types of link. The links could also be all the same size, so they can snap to copies of themselves (Fig. 9-5).

Fig. 9-3. Chain.

Fig. 9-4. Chain links.

Fig. 9-5. Single link.

One of the sprockets will be attached to power. This is the drive sprocket, and it is marked with a target in the sprocket schematic shown in Fig. 9-6. Where the drive sprocket pulls on the chain it will be tight. On the return path it has some slack and the chain will sag. You should adjust the chain length or the sprocket spacing to keep this slack small.

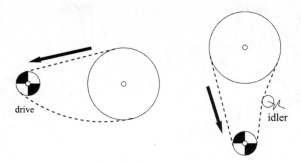

Fig. 9-6. Chain slack.

Fig. 9-7. Long chains.

The *load* is the part of the machine that is being powered. Just as an electrical load resists the flow of electricity, the mechanical load resists the driving force.

On horizontal layouts, some slack at the bottom of the loop doesn't cause any trouble. On vertical layouts you should tension the slack with a spring-loaded idler pulley (Fig. 9-6). For long layouts where the return is on the top there is a danger of the chain touching itself. That is strictly naughty, so an idler can be placed to keep the return chain out of trouble (Fig. 9-7).

Long runs don't have to be made with a single chain. The top of Fig. 9-8 shows a layout with four different chains. The three sprockets in the middle of the run are double sprockets, one sprocket for each loop attached at that point. At the bottom of Fig. 9-8 we see a single long chain that drives its sprockets in opposite directions.

The links of the chain can be adapted for different tasks. Tank treads are chains with heavy plates attached to the links (Fig. 9-9). The links can also have vertical or horizontal plates with bolt-holes, for fastening other parts onto the chain.

Toothed belts are related to chains. The belt body is made of fiber and rubber so it is strong and yet flexible. Teeth are molded into the body, and these teeth mesh with special sprockets. Timing belts in your car are built like this, and some motorcycles use a belt drive instead of a chain.

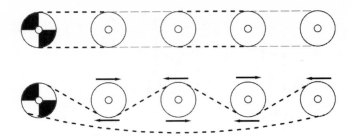

Fig. 9-8. More long chains.

Fig. 9-9. Tread.

Get rid of the teeth on your timing belt and sprockets and you have a belt and pulley arrangement. These belts tend to be wedge-shaped, fitting into a V-shaped groove in the pulley. The "V" generates more friction and keeps the belt aligned in the pulley.

A variation on the pulley and belt uses steel cable or, for really small machines, string, and wraps it around a shaft or drum. Unlike the pulley and belt, you need to wrap the cable around its support several times so that it doesn't slip.

These cable-driven systems don't have to provide continuous rotation. Sometimes you just want to drive something back and forth a little bit, such as when you bend an arm. The cable and drum arrangement works for this. You can also fasten the cables down and avoid the problem of the cable slipping. One of the attachment points can be threaded so you can tighten it as needed (Fig. 9-10). Note the bottom diagram in Fig. 9-10. The cable and idler pulley arrangement changes the axis of motion.

Cables are useful for pulling. To get a pull followed by a return, the load can have a spring to move it back to its "off" position. If you route your cable through a tube, the pull doesn't even have to be in a straight line. The cable/tube system can be snaked around corners, giving a flexible coupling

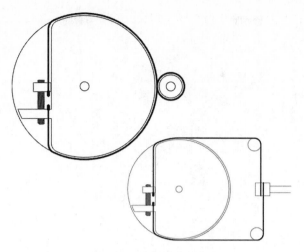

Fig. 9-10. Cable-driven joint.

Fig. 9-11. Pushrod.

between the power and the load (Fig. 9-11). The brake cables in a bicycle or motorcycle work like this.

Note that the ends of the tube need to be firmly fastened down. If the tube is fastened, the motion of the cable is transferred to the load. If the tube is left loose, the motion of the cable distorts the tube and the load is left unmoved. If the range of the motion is short and the cable is stiff, you can also push the load. Model airplanes and other radio-controlled models use this type of pushrod system.

Gears

Gears were introduced in Chapter 3 as a way to convert torque and speed. Like chains and sprockets, gears can be used to carry power (torque) from one place to another as well as to reverse the direction or axis of rotation.

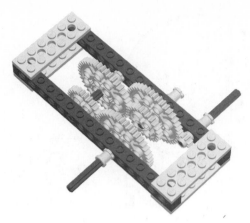

Fig. 9-12.   Gear box.

GEAR TRAINS

Try This: Normally, a box of gears is used to perform large power conversions. Figure 9-12 shows one such gear box. Note that there may be two gears on one shaft.

The input is the tiny 8-tooth gear at the right of Fig. 9-12. This meshes with a 40-tooth gear, giving a 5 to 1 reduction in rotational speed and, ignoring losses due to friction, a 5 to 1 increase in torque. A 24-tooth gear on the same shaft meshes with another neighboring 40-tooth gear, for a 24:40 or 3:5 step. This pattern repeats once more, so the output shaft shows a speed reduction of:

$$(40:8) \times (40:24) \times (40:24) = (5:1) \times (5:3) \times (5:3)$$

$$= 125:9 \tag{9-1}$$

A 125:9 ratio requires 13.9 input revolutions for one output revolution. If you used a 40:8 ratio at each of the three steps, you would have a 125:1 final ratio. That's a big reduction in speed and increase in torque.

These same gears can be stretched out to carry torque from one place to another, as shown in Fig. 9-13. In this case, there is not much adjustment from the input to the output. Each step from a small gear to a large gear slows down the motion (and increases torque), but each step from a large gear back down to a small one does the reverse. This chain is then:

$$(40:8) \times (24:40) \times (40:24) \times (24:40) \times (40:24)$$

$$= (5:1) \times (3:5) \times (5:3) \times (3:5) \times (5:3)$$

$$= 5:1 \tag{9-2}$$

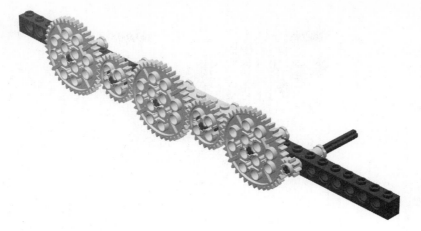

Fig. 9-13.   Gear train.

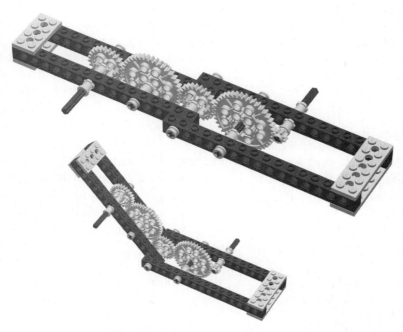

Fig. 9-14.   Gears around a corner.

Only a five times torque increase, since each 3:5 cancels out a 5:3.

Using this same type of gear train, you can carry torque around a hinged joint, as shown in Fig. 9-14. Turning the input shaft five times causes the output shaft to rotate once, no matter how the joint is bent. If you hold the input shaft steady and bend the joint, the output shaft still rotates a little bit.

Fig. 9-15. Gear elbow.

Can you figure out how far the output shaft will turn if you bend the arm 90°? Note that 90° is the same as 1/4 turn.

Lock one of the gears near the output so it can't rotate. When you turn the input, the gear train bends at the joint (Fig. 9-15). Can you find the levers in this system? How many times do you need to turn the input shaft to get the arm to bend 90°?

Backlash

Gears are more precise than pulley belts, but they still introduce errors. Since their teeth don't mesh together perfectly, there is a little bit of slack in any system of gears. You can feel this yourself using the mechanism from Fig. 9-14. Lock both the input and output shafts so they can't turn. Now carefully flex the joint. The gears, even though they don't rotate, provide enough slack to let the joint move.

This slack between the gears creates *backlash*. Backlash comes into play when the gears are moving in one direction and then reverse their rotation. The slack in the gear train lets the gears reverse without any resistance until the teeth all mesh tightly again, and then there is a jolt.

Another problem with backlash is a loss of precision. Sometimes you want to position something accurately. Any slack in your gears introduces an error into this position.

To fix these problems, gears can be preloaded. *Preloading* is where you apply tension to the gears, holding them against each other to take up any slack. One way to do this is with a *split gear*, or spring-loaded scissor gears. This is a gear that has been split into two halves, like two gears side by side on the same shaft. A strong spring is mounted between the halves so they are

Fig. 9-16.   Split gears.

forced in opposite directions. This way, the teeth are forced to mesh tightly with the connecting gear. Figure 9-16 illustrates this. As you can imagine, some trickery is needed to fasten a split gear to its shaft, and the amount of backlash protection is limited by the strength of the spring.

MORE GEARS

Try This: The gears we have studied so far have mostly been *spur gears*. Spur gears are designed to mesh edge to edge. In Chapter 8 we also saw a *crown gear* in the robot wrist machine. Crown gears are designed to mesh at a 90° angle to a spur gear, as shown in Fig. 9-17.

Another gear designed to mesh at an angle is the *bevel gear*. Bevel gears mesh with other bevel gears. Though many bevel gears are designed to work at 90° angles to each other, as shown in Fig. 9-18, the bevel can be crafted to allow other angles. Bevel gears are more efficient than crown gears and can transmit more torque.

The *worm gear* also runs at a 90° angle to a spur gear, but along a different axis. The worm gear is shaped more like a screw than a gear (Fig. 9-19). Worm gears can mesh with spur gears or gear racks. Worm gears provide a large reduction in speed and an equally large increase in torque.

Worm gears also provide another useful property. They don't back-drive easily. *Back-driving* is when you can turn the output shaft of a gear system and the input shaft turns. Worm gears tend to lock instead of back-drive.

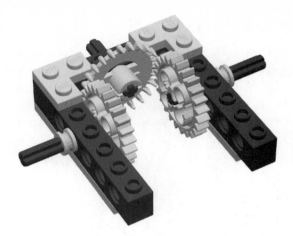

Fig. 9-17.   Crown gear.

Fig. 9-18.   Bevel gear.

Fig. 9-19.   Worm gear.

Worm Gear Driven Arm

Try This: This project expands on the worm gear base from Fig. 9-19. First, build a simple arm with a locked gear at its base as shown in Fig. 9-20. Then you can build the worm gear base and attach the arm to it (Fig. 9-21). When you turn the long input shaft the arm raises and lowers. When you stop turning, the arm stops moving. Pushing or pulling on the arm demonstrates a little bit of backlash, but no back-driving. Having a joint that locks in place means that you don't have to use power to keep it there.

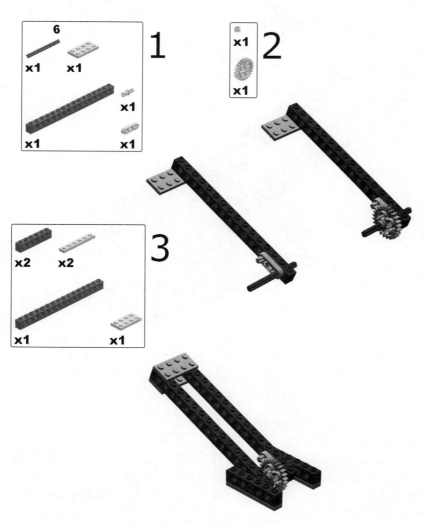

Fig. 9-20. Worm gear arm part 1.

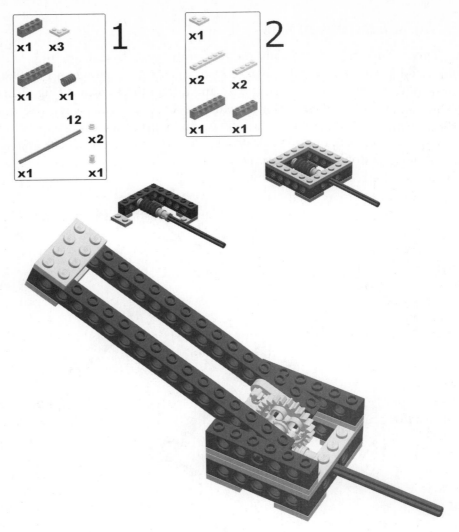

Fig. 9-21. Worm gear arm part 2.

Even More Gears

We just touched the surface of gearing systems. There are many different ways to manage power transmission through gears, sprockets, and belts. For example, you can turn one of your gears inside out and call it a ring gear. Planetary gear systems use this inside-out gear to good effect. Harmonic gears use a flexible gear to create a novel and compact gear reduction. And gearing action can be done without gears using such things as cone drives.

Couplers

Most of what we have seen in these discussions on mechanics could be described as *couplers*, that is, devices that join two parts together. In power transmission a coupler's role is to connect the power source with the actuator while adjusting for misalignments and other impediments to motion.

A coupler may be as simple as a tube with set-screws in it, used to fasten two shafts together. A step up in complexity and the tube has different-sized holes at each end to join shafts with mismatched diameters. Cut slots across the tube between the shafts and it can flex a little bit, adjusting for slight angle differences. For lightweight connections, a short length of latex tubing can work.

A spider coupler (Fig. 9-22) is like a universal joint and allows for larger differences in angle. The central cross-shaped spider fits between the input and output fittings. There are many different types of coupler available to engineers today, for all different types of torque ratings and forms of misalignment.

A different type of coupler disconnects the power from the load in the case of a jam or overload. Some attachments are naturally weak, such as a set-screw pressing on a round shaft. If the load becomes too great, the screw slips.

Using a pin or metal key to hold the shaft to its load allows more torque to be transmitted than the set-screw. By carefully sizing the key, you can arrange for it to break before your gear-train does—inconvenient, but better than ruining the whole system.

The LEGO Mindstorms set comes with a clutch, the white gear in Fig. 9-23, which is a reusable torque-limiting coupler. A *clutch* has two sliding plates that normally stick together. When the torque reaches its built-in limit

Fig. 9-22.   Spider coupler.

Fig. 9-23. Torque-limiting clutch.

the plates slip against each other. In a car with manual gearing, the clutch is opened and closed using a foot pedal.

DIRECTIONAL TRANSMISSION

Try This: A clever use of the worm gear from the LEGO set can provide power to different parts of a machine depending on which direction the motor is turning. Look at the mechanism in Fig. 9-24 to see one version of this.

When you turn the input shaft one direction the worm gear engages one of the output gears. If you reverse the input direction, the worm gear slides along the shaft and engages the other gear. This assumes that there is a little bit of a load on the shafts, at least enough to overcome the friction of the sliding worm gear.

If the output gears are on different sides of the worm gear, they both rotate in the same direction when they are engaged. If you put the output gears on the same side, they rotate in opposite directions when engaged.

DIFFERENTIAL TRANSMISSION

Try This: While the worm gear assembly in Fig. 9-24 shifts power from one output shaft to another on command, sometimes you need to shift your power in more subtle ways.

Some forms of complex couplings are called *transmissions*. They transmit power from the motor to the output, often in complex ways. Of course, gear trains and the simple couplers transmit power, too, but are instead

Fig. 9-24. Worm gear reverser.

known by their primary purpose; gearing up or down, or simply coupling. The mechanisms known as transmissions tend to provide a more intricate set of behaviors.

Look at the wheeled base in Fig. 9-25, for example, with its big wheels. Ignore for a minute how it is powered or how it might be steered. For now, you can grab it by hand and drive it around manually ("vroom" noises are optional). If it is driving in a straight line, both of the wheels turn at the same speed. If it is driving in a circle, the outside wheel moves faster than the inside wheel since it has to drive a longer distance.

If you connect the left and right wheels by a single axle, you could drive this base in a straight line, but if you tried to turn a corner the outside wheel would skid. How do you apply power to the two wheels and still be able to drive in a circle? With a differential.

A *differential* is a system of gears that applies power to two output shafts, such as our wheels, while at the same time allowing those shafts (wheels) to turn at different rates. Figure 9-26 shows a differential on this base. Note how part of it looks like the wrist gearing from Fig. 8-8.

The frame of the differential is driven by the input gear. If both wheels are carrying load and providing resistance, it drives them both equally. If one wheel is not moving, the other wheel moves twice as fast. There is always the

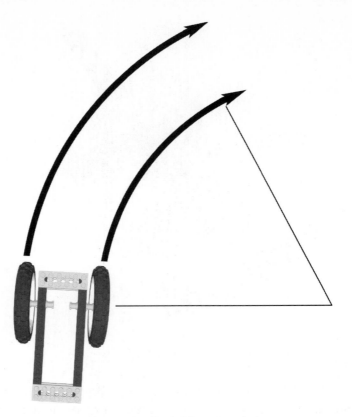

Fig. 9-25. Wheeled base (turning).

Fig. 9-26. Differential power transmission.

Fig. 9-27. Differential to measure turn angle.

same total "amount" of rotation across the two wheels. One problem with the differential is that if one wheel loses traction entirely it spins, while the wheel that is still held back by friction loses power.

You can use the differential as an input as well as an output. Instead of applying power to its frame to move the wheels, you can put a sensor on the frame and record what the wheels are doing. If both of the wheels are moving at the same rate, the frame will rotate with them. If one wheel is moving but the other is stopped, the frame rotates at one half of the moving wheel's rate.

If one wheel is moving at the same speed but in exactly the opposite direction to the other wheel, the frame doesn't move at all. To measure the turn rate of the wheeled base, you need to gear the differential to the wheels so that when they are moving in the same direction at the same rate the differential doesn't turn. To do this, we need to reverse one of the wheels with an extra gear (Fig. 9-27). Now the differential's rotation tells us how much our base is turning and in which direction.

Ratchet

We can use the differential's special ability to make our mobile base drive forward in a straight line but turn as it drives backward. One way to do this is to put a *ratchet* on one of the wheels. A ratchet lets a gear move in one direction but not the other (Fig. 9-28).

Fig. 9-28.   Ratchet on reverse.

The near wheel is invisible in this figure so you can see the ratchet. The gear replaces the spacer, and the ratchet connector meshes with its teeth. Rotating one direction, the connector rides on the teeth. Reverse it, however, and the connector locks on a gear tooth preventing further rotation. In reverse, the differential is forced to put all of the torque into the other wheel. The base turns around the locked wheel.

Summary

There are two fundamental aspects of power transmission. The first and most obvious aspect is moving mechanical force to its destination. In some special cases, this is easy: you connect the output load directly to the power source. In many cases you need to compensate for misalignment of the load and the power source. In more dramatic cases you may need to carry the power around corners, or the load could be along a different axis of rotation than the source.

In addition to moving power from one place to another, you may need to change its characteristics. Rotating force has torque and rotational velocity. Using different gearing schemes you can trade one for the other. Since motors are more efficient when they spin quickly, and most loads need power instead of speed, you use some kind of gear train in almost all mechanical applications.

By clever use of gears and other mechanisms, you can expand the behavior of your power transmission system. For example, differential gearing allows you to apply power to two wheels that may need to rotate at different speeds.

Other techniques let you apply power differently depending on whether your power source is rotating in one direction or another. A mechanical transmission can eliminate the need for electronic control systems in some cases.

Quiz

1. What do you use if your chain is getting too long or too droopy?
2. What can a chain be used for?
3. What can gears do?
4. What are split gears good for and why would you use them?
5. Name some of the gear types and describe how they are used.
6. What do couplers do?
7. What kind of gear system can apply power to two wheels while, at the same time, letting those wheels turn at different speeds?

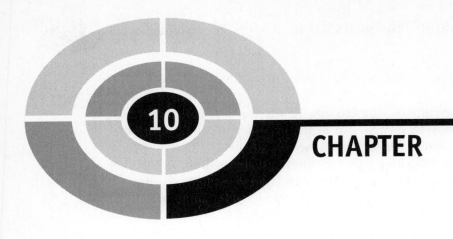

CHAPTER 10

Beyond Resistance:
Capacitance

Introduction

You may have noticed that there is a lot to learn if you want to understand robots—especially if you hope to create your own robots. Each aspect of robot creation is an entire field in itself: mechanics, electronics, control theory. Even the power drivers used to run motors have their own specialty. This chapter continues the electronics thread, adding one more piece to the puzzle.

Resistance was a simple property that we explored in Chapter 5. Resistance is pretty much a fixed attribute. Except for changes due to heating, resistors have the same effect on a circuit under all conditions.

Some components, such as the diode, change their behavior depending on the direction in which the current is moving. These components have to be plugged in correctly to work correctly.

Other components, such as capacitors, are more subtle. The effect of a capacitor is different depending on how quickly the voltage in the circuit is changing. Before we can talk about capacitance, we need to discuss the different ways voltage appears in a circuit.

AC/DC

So far, our electrons have flowed as a current from one place to another, not unlike a river flowing downstream. You might say it flows directly from one position in a circuit to another, making it *direct current* (*DC*). If you were to watch the voltage level at one point in a DC circuit, it might look like the 5-volt trace in Fig. 10-1.

Simple DC has its place, but most circuits are going to involve changes in their current and voltage. The stereotypical changing current, or *alternating current* (*AC*), is the sine wave (Fig. 10-2). The sine is the most efficient *waveform* for power transmission. The electricity in your walls is in this shape.

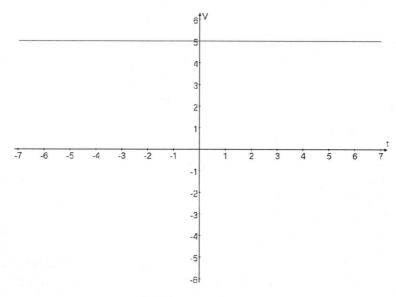

Fig. 10-1.   5-volt signal.

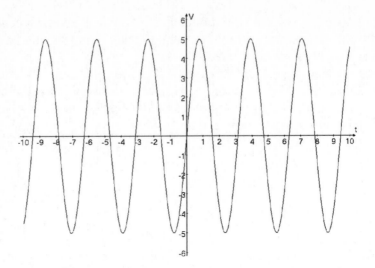

Fig. 10-2.   5-volt sinusoidal signal.

Note that though we are talking about current, voltage follows the current so a current reversal involves a voltage reversal as well. In most circuits our attention is on the voltage and not the current, so we normally show voltage measurements. There are cases where the current and voltage do not exactly match, but these subtleties don't concern us here.

Not all AC signals are *sinusoidal* (sine-like). They can be any shape. Most of the mathematics, however, describes sinusoidal signals. The sine wave is the shape that is traced out by a point on the edge of a rolling wheel, and this AC pattern is created naturally by a spinning generator.

There are several new attributes we can measure on a periodic signal: its period, for example, and its frequency and amplitude (Fig. 10-3).

It is only fair to give a math warning here. The following discussion uses a number of mathematical equations, including trigonometric functions. These equations are needed to provide a complete description. However, to get a feel for the subject you can work with the figures and the text and skim over the equations. Unfortunately for the math-phobic among us, mathematics is required for a detailed understanding of how most technologies work.

A *periodic* waveform is one that repeats itself, over and over again, coving the same voltage territory in more or less the same way over time. A *period*, then, is one *cycle* of the waveform. This period takes a certain amount of time (T).

Frequency (f) is the number of cycles that occur in one second, and is measured in *hertz* (Hz). Like most physical constants, the hertz is named after a scientist who made significant contributions to the field the constant

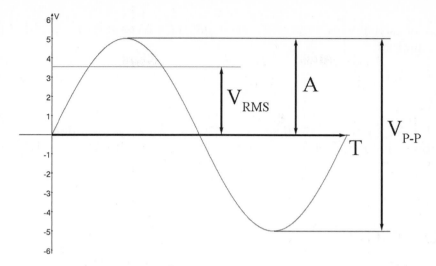

Fig. 10-3.   AC measurements.

applies to. In this case, Heinrich Rudolf Hertz, a German scientist who worked with electromagnetism and radio waves. One hertz is one cycle in one second.

A waveform's frequency and period are reciprocals:

$$f = \frac{1}{T} \qquad (10\text{-}1)$$

The *amplitude* (A) of the signal is the extreme voltage level it reaches, either above or below zero. The *peak-to-peak* amplitude is the total voltage spanned by the waveform and, for waveforms that are symmetrical around 0 V:

$$V_{P-P} = 2 \times A \qquad (10\text{-}2)$$

The voltage in our sinusoidal signal is characterized by the sine function:

$$V_t = A \times \sin(2 \times \pi \times f \times t) \qquad (10\text{-}3)$$

The magic number $(2 \times \pi)$ is the number of radians spanned by one circle. A *radian* is a measurement of angle, like the more familiar *degree*. Where there are 360 degrees in a circle, there are $(2 \times \pi)$ radians. Therefore, there are $(2 \times \pi \times f)$ cycles per time unit t.

There is one last question. How much power is represented by this waveform? For a DC signal, power is the voltage times the current. But for AC the voltage is constantly changing. To calculate the average power for a sinusoidal signal we use the *root mean square* (*RMS*) amplitude instead of the raw amplitude:

$$V_{\text{RMS}} = \frac{A}{\sqrt{2}}$$

Our 5 V AC signal is equivalent to a 3.5 V DC signal. If we wanted to match the power of our 5 V DC signal we would have to generate a sinusoidal signal that has 7 V peaks.

OSCILLOSCOPE

Measuring resistance is done with an inexpensive multimeter, like the one shown in Chapter 5. That same meter does a fine job of measuring DC voltage and AC RMS voltage. What it won't do is give a picture of the voltage levels over time. A far more expensive tool does this, the *oscilloscope* (Fig. 10-4).

Where calculated diagrams like Figs. 10-2 and 10-3 can show us the math and the theory of the circuit, oscilloscopes show us a representation of what is actually happening. They are invaluable tools for the electronics workbench, with their only downside being their price. A basic oscilloscope like the one pictured in Fig. 10-4 costs $400 or so, with the better models starting at $1,000 and the best models costing as much as a small car.

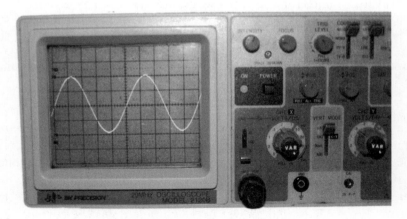

Fig. 10-4.   Oscilloscope.

Diodes

SIGNAL DIODE

The *diode* manipulates a circuit's voltage. While there are a number of nuances to the diode's behavior, it has two primary effects on circuits.

First is its unusual ability to conduct electricity in only one direction. Of course, if you apply too much voltage to a diode it will, like most components, break down and stop working correctly.

Note the arrow in the diode's schematic figure in Fig. 10-5. This is the direction of the *forward current* (illustrated in terms of conventional current), flowing from the anode (A) to the cathode (C). The cathode line on the diode itself matches the line on the schematic. You can say that the diode's arrow points toward ground, or the circuit's negative terminal. For example, in the circuit in Fig. 10-6, the diode allows electricity to conduct through the resistive load.

The second attribute of the diode is its forward voltage drop. The voltage leaving a diode is always less than the voltage entering the diode. It is important to point out that, in spite of this, a diode doesn't have a significant resistance. This voltage drop is typically less than a volt, often about a half-volt drop. In some cases this is just an annoying side effect. However, you can also design circuits to specifically use this voltage drop.

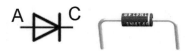

Fig. 10-5. Diode.

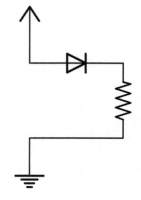

Fig. 10-6. Diode circuit.

There are many other subtle attributes of diodes, such as how quickly they can respond to changing voltages, how much they "leak" when reverse current is applied, their breakdown voltage beyond which they burn out, and others.

RECTIFIER

If you pass alternating current through a diode it does not emerge unchanged on the other side (Fig. 10-7).

A *rectifier* converts AC to DC, more or less. One diode is a half-wave rectifier, since only half of the waveform makes it through the diode. The circle with the sine wave in it is a generic signal generator and is the source of our AC signal. The tap in the upper left of the circuit shows where we measure the signal voltage relative to ground. Likewise, the tap after the diode shows the effect the diode has on the signal. The resistor is a generic load and isn't really relevant to the rectification process.

Half of our power is being blocked by the diode. If you wanted to keep the signal positive without losing half of its power, you need more diodes to make a full-wave or *bridge rectifier* as shown in Fig. 10-8.

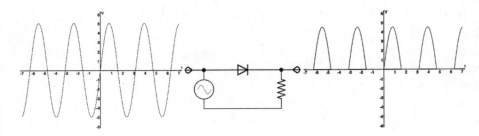

Fig. 10-7.   Half-wave rectifier.

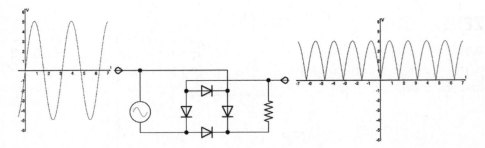

Fig. 10-8.   Full-wave rectifier.

The rectifier is an important part of all circuits that convert AC signals into DC. The signal coming out of the rectifier is still AC, but it is at least all on the same side of ground level. What path does the power take through the bridge rectifier when the input signal is positive? What path does the negative signal take? Remember that conventional current follows the direction of the diode's arrows.

Try This: Take four diodes and assemble them in the rectifier configuration. Now hook up your 9 V battery to the inputs and measure the output. Reverse the battery input and see if the output changes at all. Does it? What voltage do you get out of the rectifier? Is it the same in both configurations? Measure the battery to see how much you are putting in.

The next step in the AC to DC signal conversion will be filtering. Capacitors, conveniently enough, can be used to filter and condition AC signals.

LIGHT-EMITTING DIODE

Another useful side effect of the diode is that, in the correct form, it will emit light. Light-emitting diodes (LEDs) are used everywhere in electronic devices to signal and communicate to the user.

Before LEDs were developed, electronic devices had to use ordinary, though very small, resistive light bulbs. LEDs use less power than lights and, as their price dropped to their current low levels, rapidly took over their job. Now the white ultra-bright LEDs are even replacing light bulbs in flashlights, streetlights, vehicle signal indicators, and other high-intensity applications.

An even more intense form of light-emitting diode is the laser diode. This is used in all of your CD and DVD players as well as in industry and science. It also makes a great cat toy.

ZENER DIODE

While most diodes are used in the forward conduction mode, the Zener diode performs its function in the reverse mode. The *Zener diode* (Fig. 10-9) was first described by the American physicist Clarence Melvin Zener. When the

Fig. 10-9.   Zener diode.

input exceeds the breakdown voltage in regular diodes, the diode is destroyed. The Zener diode, however, has a very low breakdown voltage and can conduct power in reverse mode without being damaged. Up to a limit, of course.

In the forward mode the Zener diode acts like a signal diode. In reverse mode it blocks current like the regular diode until the voltage gets high enough and then it conducts, unlike the regular diode. This effect can be used to limit input voltages and keep circuits safe from overloading. The Zener diode in conjunction with a signal diode can clamp an input, forcing it to remain between zero and the Zener diode's breakdown voltage (Fig. 10-10).

If the Zener is set to conduct at a reverse of 4 V, the output of the circuit in Fig. 10-10 might look like Fig. 10-11. There are many other clever ways that diodes can be used to modify or limit a signal. See if you can think of a few more yourself.

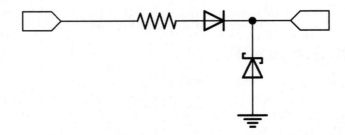

Fig. 10-10. Diode clamping circuit.

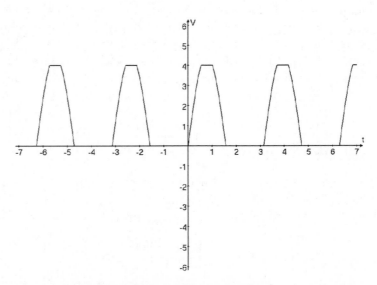

Fig. 10-11. Clamped signal.

Capacitors

CAPACITOR

Everything acts like a capacitor to some extent, the same as everything is a resistor. Of course, there is also a specific component called a *capacitor* that provides a specific capacitance. Capacitance is measured in *farads*, named after Michael Faraday, and is represented by the letter *F*. Michael Faraday was training to be a bookbinder when he developed an interest in science. From there he went on to do many things, including laying the foundations for our understanding of electromagnetism.

The symbols for the capacitor are shown in Fig. 10-12, as well as pictures of a couple of representative components. We'll return to this figure shortly.

While it is easy to get a feel for what a resistor does, capacitors can be a bit tricky. Let's start with some history.

Condenser

The capacitor was originally called a *condenser* and it consisted of a glass jar with metal foil wrapped around its inside and outside. Wire electrodes were attached to the foil and then to a generator or other source of electricity. You could, for example, scuff your feet on a carpet and touch one of the electrodes to charge the condenser.

This glass jar condenser is called a *Leyden jar*, and was invented by Pieter van Musschenbroek from the University of Leyden in the Netherlands, and later improved by the English scientist William Watson. It was called a condenser because people thought that electricity was a fluid and that this device condensed it. What it does is store an electric charge.

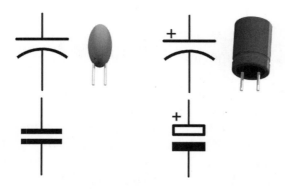

Fig. 10-12. Capacitor.

Leyden jars were used to store *static electricity*. Static electricity is high-voltage electricity. Static electricity shows up as a charge imbalance on the surface of an object and can even be collected on insulators. Once a charge is applied to one of its electrodes, the condenser remains charged until it is discharged. This voltage could leak away, carried off by little bits of moisture in the air, or it can be applied to a circuit. Touching both terminals together can create a spark.

While it is easy to make, the Leyden jar can also be dangerous. Variations of this device are constructed for high-voltage equipment, such as Tesla coils.

Modern construction

The modern capacitor is essentially the same as the Leyden jar, on a smaller scale.

There are three parts to a capacitor. In the center there is an insulator, called the *dielectric*, meaning "stuff that won't conduct electricity." A dielectric is transparent to electrical effects, such as the electric field, but not the electrons themselves.

On either side of the insulator is a thin *electrode*. An electrode is the point where electricity enters or leaves an electronic device, and all components have two or more.

Attached to each electrode is a terminal wire which makes it easy to hook the capacitor into a circuit. Note how the schematic symbol in Fig. 10-12 reflects this internal construction.

The electrodes and dielectric may be wrapped up into a tight cylinder. The inside and outside of this spiral electrode need to be separated by more dielectric so they don't short out.

Behavior

Now imagine the capacitor in a circuit such as the one in Fig. 10-13. At the start, there is a surplus of electrons on the battery side of the switch, crowded together and not happy about it. When the switch is pressed they are suddenly connected to the electrons on the other side, and promptly start to push them around. The electron pressure is increased all around the switch like the big shopping day after Thanksgiving.

The capacitor, however, does not let the electrons through. The electrons in the pressurized parts of the circuit, including one electrode in the capacitor, are under a certain amount of pressure. They crowd together more

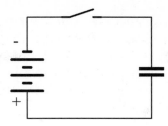

Fig. 10-13.   Capacitor with charging switch.

than they would normally, and this increases the electric field along the wires and, of interest here, at one of the electrodes in the capacitor.

This electric field has influence through the dielectric and, though the electrons can't slip through, the electric field pushes against the field of the electrons on the other side. This, in turn, pushes those electrons away from the dielectric a little bit.

In short, once the switch is pressed, there is a small jolt of current across the capacitor and then it all stops. If the switch is released the capacitor keeps its charge, so additional button pushes don't do anything until the capacitor is neutralized somehow, draining excess electrons from the crowded side of the dielectric.

The result of this behavior is that the capacitor appears to conduct electricity while the voltage is changing, but does not conduct when the voltage is steady. If you apply an AC signal on one side of a capacitor, particularly one that goes negative as well as positive, a version of that signal will appear on the other side of the capacitor.

The capacitor has a capacity, its capacitance as described in farads. This capacity determines how much of a charge it can hold. The capacitance, in conjunction with the frequency of the incoming signal, determines how efficiently that signal is transferred through the capacitor.

There is one other detail. While many capacitors can be wired into a circuit in either direction, like a resistor, some capacitors have one terminal marked as positive. These *polarized* or *electrolytic* capacitors only work in one orientation. Wire them in backwards and they can fail, sometimes dramatically.

Charge

First we must note that one farad is a lot of capacitance. You rarely use capacitors that large. In most electronic circuits you work with capacitances in the range of the microfarad, which is marked as "μf". As we saw in Table 4-1, the Greek letter *mu* (μ) stands for 0.0000001, which is fairly small. Capacitors in the nanofarad (nf) and even picofarad (pf) range are not uncommon.

The capacitor stores charge, which is measured in coulombs. In this context, the coulomb charge is represented by the letter Q, since C stands for the capacitance as measured in farads. V is still voltage, in volts:

$$Q = C \times V \qquad\qquad (10\text{-}4)$$

One electrode on a capacitor can have a charge of Q, while the opposite electrode has a matching charge of $-Q$. The total charge is zero. The electrons are just crowded on one side and sparse on the other.

Another way to look at this is with a water analogy. A battery is like a large reservoir of water (Fig. 4-3), and a resistor is like a narrow spot in a connecting pipe (Fig. 10-14). A capacitor, then, is like a small reservoir with a rubber diaphragm across the middle (Fig. 10-15). Water can't flow across it, but pressure on one side can push water out of the other side.

We explore the details of how the capacitor reacts to an AC signal in the section on capacitor networks. First, there is an experiment you can do to hear for yourself how a capacitor modifies an AC signal—if you dare.

CAPACITORS AND AUDIO

Try This: It can be difficult to see how a capacitor affects your circuit. First you need a source of an AC signal and then you need a way to watch this signal. Normally, electronics engineers use a signal generator to create a controlled signal and then an oscilloscope to read it back. With these tools it is easy to see what your circuit is doing. You, however, may not have these expensive tools on hand. You may, however, have two other sophisticated tools that you can use with the other parts in Table 10-1.

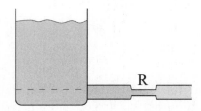

Fig. 10-14.   Water analogy for resistors.

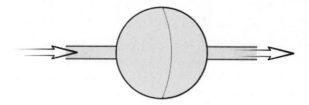

Fig. 10-15. Water analogy for capacitors.

For a signal generator, you can use your computer or even your stereo or portable music player. For the detector, biology has given you a fine pair of ears. With any luck, both of these are still in good working condition.

If you have a computer you may also have an extra pair of cheap speakers. If you have a stereo system, the speakers that plug into it may have bare wire for their hookup. Either way you need a way to get access to the wire that runs between the audio generator and the output speaker.

I found the cheap speakers that came with my computer and cut and stripped the wire on one of the speakers. Before you start cutting wires around your house, be sure you have permission to do so. Spouses and/or parents can be very particular about the shape of their electrical appliances.

Once you have access to the wires, it is a simple task to wire a capacitor to them. Turn on a song that has a lot of high and low notes. A squeaky singer against a thrumming backdrop of bass and drums would work. Use your own judgment. With the speaker still attached, listen to the music to get a sense of how it sounds. Now cut those wires and wire capacitors in series and parallel as shown in Fig. 10-16.

Table 10-1 Parts: capacitors.

Parts	
C	Capacitors: 0.1 µf, 1 µf, 10 µf, etc.
Tools	
Speaker	
Speaker wire (probably already attached to the speaker)	
Computer, stereo, boombox, or other player	
Working pair of ears	
Soldering iron	

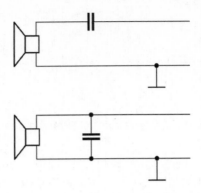

Fig. 10-16.   Capacitor and your speaker.

Many speaker wires have a bare wire wrapped around an insulated central wire. This bare wire is typically the ground wire. When wiring the capacitor in series, use the signal wire. If you can't tell which is which, try both.

Use different capacitor sizes, such a 0.1 µf, 1 µf, 10 µf, and even larger if you can find them. The exact values don't matter, so a 4.7 µf is just as interesting as a 1 µf capacitor. Note how the smaller capacitors let higher frequencies through, while the larger capacitors favor the lower frequencies. Note also the difference when the capacitor is in parallel.

When you are done, you should put everything back together the way you found it. Twist the wires together if you cut them and, for bonus points, solder them together. Use electrical tape to insulate the junction and it's as good as new, though twice as ugly.

Capacitor Networks

Resistors had rules to describe the behavior of two or more resistors in series or in parallel. Capacitors have similar rules. In fact, the rules are the same but the meanings are reversed.

For two or more capacitors in series (Fig. 10-17) the capacitance is reduced:

$$C = \frac{1}{\dfrac{1}{C1} + \dfrac{1}{C2} \cdots \dfrac{1}{CN}} \tag{10-5}$$

For two capacitors, this reduces to:

$$C = \frac{C1 \times C2}{C1 + C2} \tag{10-6}$$

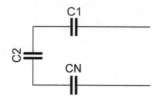

Fig. 10-17. Series capacitors.

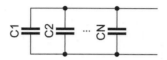

Fig. 10-18. Parallel capacitors.

Capacitors in parallel (Fig. 10-18) increase the capacitance:

$$C = C1 + C2 + C3 \qquad (10\text{-}7)$$

RC CIRCUITS

Remember as we run through these circuits that when you apply voltage to a capacitor, it charges up while at the same time pushing on the electrons across the dielectric. When that voltage is removed, the capacitor holds the charge. If a circuit is then made between the capacitor's electrodes, the electrons on both sides of the dielectric try to achieve balance and the capacitor discharges.

Note that none of these operations are instantaneous. Everything takes time. And this time delay is an important part of how we use the capacitor in electronic circuits.

A resistor added to a capacitor slows down the capacitor's reaction. Together they are an RC network. Carefully choosing the values for the resistor and capacitor, you can fine-tune the network so that it responds to the desired frequencies.

Charging and discharging

RC circuits are characterized by a *time constant*. This time constant is represented by the Greek letter τ (tau):

$$\tau = R \times C \qquad (10\text{-}8)$$

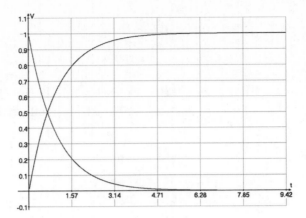

Fig. 10-19. Capacitor voltage over time.

Resistance R is in ohms and capacitance C is in farads, with time in seconds. A circuit with a $1\,k\Omega$ resistor and a $1\,mf$ capacitor has a time constant of 1 second. The time constant represents the amount of time it takes for the RC circuit to charge or discharge. If the capacitor is full, it takes $(2 \times \pi \times \tau)$ seconds to charge or discharge (Fig. 10-19).

The *cutoff frequency* of an RC circuit is described by equation (10-9). The cutoff frequency is the frequency where the RC circuit stops being effective. What this means depends on the type of circuit.

$$f_c = \frac{1}{2 \times \pi \times R \times C} \tag{10-9}$$

This relationship of time to frequency was first seen at the top of this chapter in equation (10-1).

A charged capacitor discharges its voltage V_0 over time according to equation (10-10), which is reflected in the falling curve of Fig. 10-19:

$$V_t = \frac{V_0}{e^{t/(R \times C)}} \tag{10-10}$$

The voltage over time of an RC circuit being charged with a voltage V_{in} is that same equation turned upside-down:

$$V_t = V_{in} - \frac{V_{in}}{e^{t/(R \times C)}} \tag{10-11}$$

RC FILTERS

RC networks can make simple *filters*. A filter is a circuit that lets through some AC signals but not others. Filters vary in their effectiveness. Simple filters like the ones shown here pass the target frequency but also a bunch of others; they work, but don't have a very sharp *frequency response*.

A *first-order filter* halves the output signal strength for every doubling of the input frequency away from the cutoff frequency. A better filter is the *second-order filter* which reduces the output strength by 3/4 for every doubling. *Third-* and *fourth-order filters* are even more effective.

High-pass filter

A *high-pass filter* allows high-frequency signals through but reduces the strength of low-frequency signals below the filter's cutoff frequency. The cut-off frequency is described by equation (10-9).

The circuit in Fig. 10-20 is a simple high-pass filter. The input signal at the pad on the left is, optimally, a sine wave for the math to work correctly. Different waveforms will behave differently. We can watch the transformed signal at the test point on the right side of the circuit.

An interesting thing to note is that the load that we are driving is the resistor in the RC circuit. If your load has a huge resistance, such that it doesn't really "exist" in terms of the circuit, you need to add a smaller resistor across it in parallel to provide the R in the RC time constant.

The frequency response of this circuit, given a sine input, is:

$$A = \frac{1}{\sqrt{1 + (f_0/f)^2}} \tag{10-12}$$

where f_0 is the cutoff frequency, f is the input frequency, and A is the amplitude *gain*. Gain is the amount of *amplification* or, in this case, *attenuation* a circuit provides. Amplification increases a signal, with a gain greater than 1. Attenuation decreases a signal, with a gain less than 1.

Our high-pass response curve is shown in Fig. 10-21 for an f_0 of 1 kHz. Note that the time axis is marked in units of 1 kHz. For low frequencies the output is greatly reduced, while for frequencies above our target of 1 kHz most of the signal gets through.

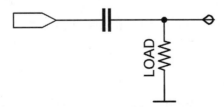

Fig. 10-20. High-pass RC filter.

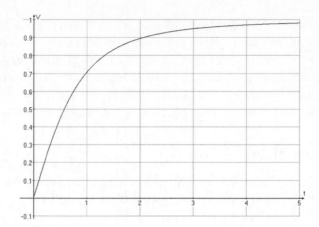

Fig. 10-21. High-pass frequency response.

Low-pass filter

A *low-pass filter* is the reverse of a high-pass filter. In this case, we want the lower frequencies to get through while blocking higher frequencies. A simple low-pass filter is shown in Fig. 10-22. If the load is high-impedance, this filter is described by the simple equations listed below. However, be aware that any circuit resistance downstream from the circuit can affect its behavior.

Everything links together. Even though any one piece of a circuit may be simple, all of the pieces working together can be complex and hard to understand. This system complexity is found in any complicated field of study.

We threw in a new word, *impedance*, which has a symbol Z and is measured in ohms. For a simple DC circuit, impedance is essentially the same as resistance. For AC circuits, it is more complex. In general, impedance is the measure of how a circuit resists, or impedes, current.

The frequency response for the circuit in Fig. 10-22 is:

$$A = \frac{1}{\sqrt{1 + (f/f_0)^2}} \tag{10-13}$$

This is illustrated in Fig. 10-23 for an f_0 of 1 kHz.

Other filters

There are two other common filters. The *band-pass filter* is like a low-pass filter followed by a high-pass filter (Fig. 10-24). A *band-gap* or *notch filter*

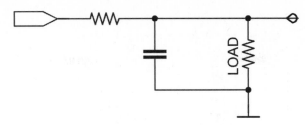

Fig. 10-22.   Low-pass RC filter.

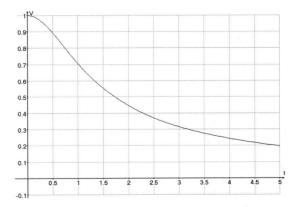

Fig. 10-23.   Low-pass frequency response.

Low-Pass   High-Pass

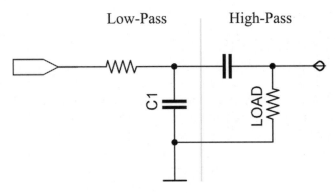

Fig. 10-24.   Band-pass RC filter.

is a little more complicated, allowing frequencies above and below the cutoff through but limiting signals at and near the cutoff. This is like a high-pass filter in parallel with a low-pass filter (Fig. 10-25). For both of these filters you need to select the high- and low-pass cutoff frequencies so that the overlap is correct for the filter shape you want.

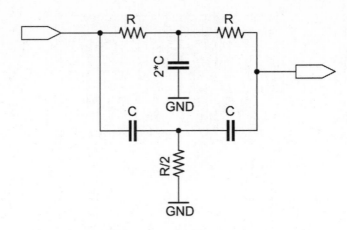

Fig. 10-25.   Twin-T notch filter.

Audio filter

Try This: Design some of these filters, and then build them on your bread-board. Plug them into the audio circuit you used for the capacitor test and see what they do. The load resistor is the speaker, and this will be 4 or 8 ohms. If it is an amplified speaker then all bets are off. Change the component values and see how they affect the filter.

DIODE-CAPACITOR CIRCUITS

Power filter

Some interesting effects can be had by mixing capacitors with diodes. For example, we can revisit our full-bridge rectifier from Fig. 10-8 and add a large capacitor to it to smooth out the *ripple*. Ripple is the amount by which a signal wobbles away from a pure DC line. Figure 10-26 shows a filtered AC to DC power circuit. This is a very common circuit, found in all sorts of systems.

You want a large capacitor in this circuit. You aren't tuning the circuit against a specific frequency so much as you are using the capacitor's ability to store an electrical charge. The signal charges the capacitor as the voltage rises so the output is somewhat less than it would normally be. As the signal voltage drops the capacitor discharges, raising the output voltage above what it would normally be. The result is to smooth out ripples in the signal.

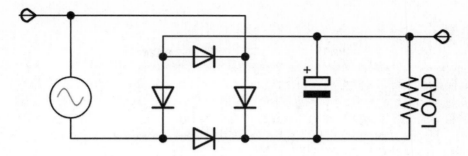

Fig. 10-26. Power filter.

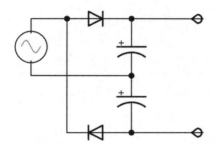

Fig. 10-27. Voltage doubler.

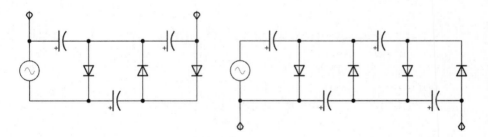

Fig. 10-28. Voltage multipliers.

Voltage multiplier

A circuit that seems like magic is the voltage multiplier. I present it here as a bonus circuit. Try it out. Try to figure it out.

Figure 10-27 is a simple voltage doubler. This circuit, like the others, requires an AC signal on the input to work correctly. Figure 10-28 shows a voltage tripler and quadrupler. More diode networks can be added to get as much voltage as you can stand.

Summary

There is a lot of difference between the simple world of DC and the rich dynamics of AC signals. After a brief introduction into the basics of sinusoidal signals we went and looked at ways of turning AC back into DC.

The diode is our first component that has to be inserted into a circuit in the correct direction. This is because they only conduct electricity in one direction. Except, of course, for Zener diodes. These two types of diodes give us the ability to control and limit input signals.

The main focus of this chapter was the capacitor. While capacitors block direct current, they allow the electric field to pass through and move the electrons on the other side of their dielectric.

Capacitors are normally used in conjunction with resistors to make filters, through they have their uses in other types of power circuits. The RC time constant is a common thread that runs through a variety of capacitor/ resistor circuits and defines how quickly that circuit can react to changing voltages.

Quiz

1. What type of electric current comes out of your wall? What is the other type of current?
2. What is frequency?
3. What does a diode do? What happens when you put one in a circuit backwards? What happens if you apply more voltage to a diode that is plugged in backwards? To a Zener diode?
4. What are the parts of a capacitor?
5. Does a capacitor conduct electric current? What does a capacitor do?
6. What is a common circuit using a capacitor and a resistor? What is the formula that describes its behavior?

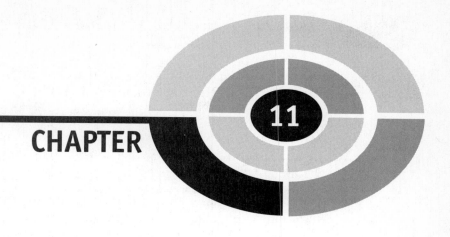

CHAPTER

11

Inductance and Magnetism

Introduction

An inductor is the inverse of a capacitor. Where the capacitor blocks DC and passes AC, the inductor blocks AC and passes DC. The capacitor uses the electric field for its effect while the inductor uses the magnetic field. But we are getting ahead of ourselves.

Magnetism is the other half of the electromagnetic field introduced in Chapter 4. The capacitor manipulated the electric field and now we get to play with the magnetic aspect.

Magnetism seems magical in its effects, pushing and pulling at metal from a distance. While there are different types of magnets, we look at the magnetism that is created by a moving electric field. Not only can we use magnetism to create physical forces to drive our robots, we can use it to affect the behavior of our electronic circuits.

Electromagnets

An *electromagnet* is a temporary magnet created by current in a coil of wire.

Recall for a moment the perpendicular forces of magnetism, electric current, and physical force from Figs. 4-1 and 4-2. Whenever you run electricity through a wire, some magnetism is generated. Normally, though, this is not enough to be useful. If you put enough current through a wire you might get some noticeable magnetism, but increasing the current to that level is rarely practical.

What you need is a way to gather a whole bunch of magnetism together, collecting it into a bundle, until it has a useful strength. A tight coil of wire creates this bundling of magnetic fields. While electrons are flowing through it, the coil is a magnet with a North and a South pole and magnetic flux looping between the two (Fig. 11-1). *Magnetic flux* is, roughly, the stuff of magnetism as defined by how dense the magnetic field is over a particular area of space. The magnetic field is normally shown as lines curving through

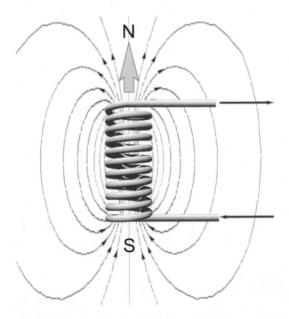

Fig. 11-1.   Air-core electromagnet.

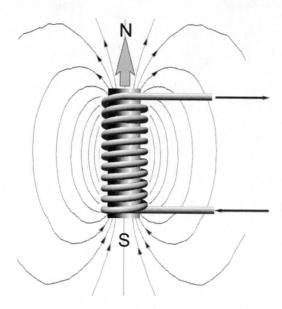

Fig. 11-2. Iron-core electromagnet.

space. The closer together the lines are, the denser the flux, and the stronger the magnetic field at that point.

While a coil of wire does a good job of creating a usable magnetic field, adding a soft iron bar to the center of the coil means the field will be much stronger (Fig. 11-2). In fact, an iron core can make the field hundreds or even thousands of times stronger. The electrically induced magnetic field causes the otherwise randomly oriented magnetic fields within the iron, known as *magnetic domains*, to line up and add to the strength of the coil.

NAIL ELECTROMAGNET

Try This: The usual experiment in electromagnetism is to take a large nail from the hardware store and wrap a lot of fine, insulated wire around it. Try it! Use 100 wraps or more. If you can find lacquer-covered "bell wire," that's great, or you can use the thinnest wire you can find at the electronics shop. The more wire you use, the stronger the magnet will be. Once wrapped, hook the ends of the wire to a battery and the nail becomes a magnet. Until you remove the battery.

Another reason for a long wire is resistance. The longer the wire the more resistance it has, and the better it is for the battery. Short, hot wires can be hard on both the battery and your fingers.

RELAY

An electromagnet is not just good for picking up paper clips and iron filings. You can put an electromagnet inside of a switch, for example, to make an automatic switch or *relay*.

As in all things electronic, there are schematic forms for everything. In this case, wire wrapped around a core looks like a loopy or wiggly line with a bar next to it (Fig. 11-3, left-hand relay). The symbol for the relay is an electromagnet above a switch. The electromagnet symbol is often abbreviated to a simple box with a diagonal line through it, as shown in the right-hand relay in Fig. 11-3.

MOTORS

Figure 4-2 in Chapter 4 showed the interaction of forces around a wire carrying a current. In that figure, the magnetic flux is flowing into the page, the electrical current is moving from right to left, and the wire is being pushed down. Figure 11-4 illustrates the same idea showing the physical magnets. Note that the magnetic field points from the North pole to the South pole. As current is applied to the wire it generates its own field which interacts with the flux around it, pushing the wire up.

If you make your wire into a loop and run it back and forth between the magnets, the forces push up on one side and down on the other side of the loop (Fig. 11-5), creating a turning force or torque on the wire. Finally, instead of a single wire you can create coils of wire around a soft iron core

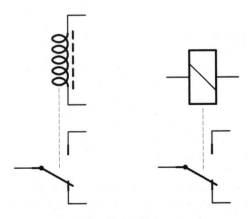

Fig. 11-3. Relay.

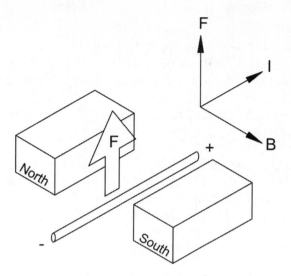

Fig. 11-4. Magnetic forces.

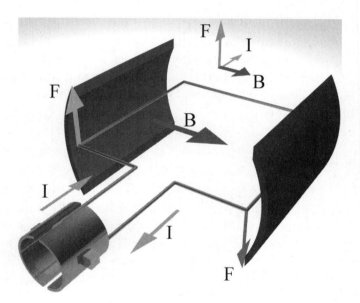

Fig. 11-5. Basic motor.

and the field increases hundreds or thousands of times and you get some serious force out of your motor.

That is what you will see if you look inside a motor. The ring of permanent magnets around the outside of the motor's case is called the *stator*, because it is static (it doesn't move). The stronger these magnets are, the more powerful your motor can be. The wire-wrapped loops and their metal core that spins inside of those magnets is called the *rotor*, because it rotates.

Electricity is applied through a rotating switch called the *commutator*, because it helps the electricity commute, or travel. The electrodes of this switch are spring-loaded contacts or *brushes*. When electricity is applied to the coil, the coil's magnetic field pushes against the static field and makes the rotor turn until the rotor's magnetic field is back in line with the stator's magnetic field. And then it stops.

This is where the commutator comes in. As the rotor turns, the commutator acts like a DPDT switch and reverses the current through the coils just as they are coming in line with the stator's field. This reverses the coil's field, pushing the rotor against the new magnetic alignment and the process repeats until the motor wears out.

GENERATORS

When you put electricity into a motor it spins. What happens if you grab the rotor of a motor and turn it by hand? Electricity comes out. This, then, is a generator. Motors and generators are essentially the same thing. There are subtle differences in construction that make motors more efficient at creating torque and generators more efficient at creating electricity, but their basic construction is the same.

Magic motors

Try This: If you hook two motors together, when you turn one of them by hand, the other one turns by magic (Fig. 11-6). Well, by electricity, though Sir Arthur C. Clark once said that *any sufficiently advanced technology is indistinguishable from magic*. The difference is in how well we understand what we are seeing.

SERVOS AND STEPPERS

The system shown in Fig. 11-6 is a form of *servomechanism*, or *servo*. A servo is a system that provides controlled motion at a distance, typically using feedback. The generator/motor pair is crude, but the output motor does mirror any motion from the input generator. A true servo system would mirror the input position even in the face of resistance, using an internal

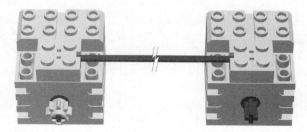

Fig. 11-6.   Motor as generator.

feedback loop to measure the actual position and update the mechanism as needed.

Radio-controlled vehicles of all kinds rely on servos to control their steering and throttle. These R/C servos consist of a tiny motor and gear train, a feedback sensor on the output shaft, and a circuit to control it all. The servo is given power and a control signal, and the circuit uses feedback from the position sensor to keep the output shaft at the commanded position.

Servos are handy because they move the feedback circuit into the motor itself. This is like taking a control function out of the brain and putting it into a neural reflex. The reflex loop is shorter and faster, and it can happen without having to bother the higher-level systems.

A stepper motor is a motor that moves in short, controlled steps instead of in a continuous spin. In a continuous motor, the commutator reverses the current to provide constant rotation. A stepper motor doesn't do that for you. An external circuit is needed to reverse the fields and cause the next step. Stepper motors are good for accurately positioning the motor and then holding that position. Stepper motors are often used in robots because of this fine level of control.

Inductors

This exploration of magnetism was interesting in its own right, though the point was to see how electricity and the magnetic field interact.

An inductor has some aspects of a motor and some aspects of a generator, with none of the moving parts. Inductors are constructed just like electromagnets, with wire wrapped around a core. Instead of soft iron, however, inductors tend to use *ferrite* cores. Ferrite is a type of ceramic filled with iron, boron, or some other metal that reacts with magnetic fields. Ferrite can store a stronger magnetic field than plain iron, giving you more power for a given core size. Ferrite is also brittle, so it isn't used in motors and generators.

BEHAVIOR

Let's look at what happens when we apply a charge to a simple wire, creating a voltage differential across the wire. But let's slow down the action a lot.

When you first apply a charge to one end of the wire it creates a charge imbalance, causing the electrons to move along the wire. This, of course, is current.

A change in current creates a magnetic field that grows outward from the wire. This is a moving magnetic field and, as discussed in Chapter 4, a magnetic field moving across a wire creates current. This generated current opposes the applied current, reducing the actual current flowing through the wire.

Once the current has stabilized, the magnetic field stops moving. When the magnetic field stops moving, it stops generating the opposing current. At that point, the wire is just a wire again with its ordinary resistance and a stable current.

You may have noticed that the current in the wire creates a magnetic field that grows across the same wire *that created the field*, in turn generating a current *opposite* to the one that started the process. And you would be correct.

What happens when you stop the current in the wire? The magnetic field moves again, in the opposite direction, as it collapses. This creates a current working to keep the flow moving. When the magnetic field is gone, that generated current fades away with it.

Inductors are like flywheels for current. They don't like to start moving and, once moving, they don't like to stop. Another way to look at the inductor is as a magnetic reservoir. When you first apply current to an inductor, it takes that energy and uses it to fill up its magnetic field. When the current is removed, the inductor empties its magnetic field, converting it back into a current.

COMPONENT

This opposition to change is called *inductance* and all wires and components in your circuit are, to some extent, *inductors*.

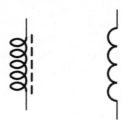

Fig. 11-7.   Inductor.

If you wrap a wire around a ferrite core, it has a lot more inductance than a single wire in space. This would be an inductor as a component, as shown in Fig. 11-7. Inductance is measured in the *henry* (H), after the American scientist Joseph Henry who discovered a number of the phenomena that we call inductance. Inductors are often represented by the letter *L* in formulas and schematics. As in capacitors, we will normally work with microhenries (μH) and millihenries (mH) instead of full henries.

You could experiment with the inductor the same way as you did with the capacitor, by inserting it into an audio circuit and listening to the changes. Where the capacitor prefers to pass high-frequency signals, the inductor passes low-frequency signals.

FILTERS

Try This: Since the inductor's response changes with the frequency of the signal, it can be used in a resistor-inductor (RL) filter, replacing the capacitor in the RC filter. Remember, however, that the inductor's response is the reverse of the capacitor's response.

You can also combine the inductor and capacitor to make filters. Put an inductor by itself in series with your audio test station and see how it affects the signal. Try building filters with the inductor, replacing the capacitor in your previous circuits, and see how they act.

PHASE

We are ignoring one aspect of capacitors and inductors: how they transmit voltage and current at different times in the AC cycle. When the voltage from a circuit peaks at a different time than the current, there is a *phase change* between them.

Now that you know this phase change exists, it won't surprise you when you find it in other electronics books. But we aren't going to explore it here.

TRANSFORMER

An inductor works with one coil of wire around a core. What if you put two separate coils of wire around the same core (Fig. 11-8)?

When you apply voltage across one of the coils, generating a current in the coil, it creates a magnetic field. This field creates the self-inductance discussed above. As the magnetic field grows, however, it also crosses the wire in the second coil. This moving field creates a current in that coil.

A changing current in one coil creates a copy of itself in the other coil. The two coils don't even have to be wrapped around the same part of the core. The magnetic field fills the entire core, even the bits that extend outside of the coil. In fact, many transformers look like the one illustrated in Fig. 11-9.

The obvious experiment to try is to change the number of loops on one side of the transformer. What happens then?

The transformer acts like an electromagnetic gear. When the coils have the same number of loops, the output matches the input in terms of current and voltage potential. If there are more input loops, or windings, than output windings, the magnetic flux has to transfer the energy of the generating coil, called the *primary*, to a smaller output coil, called the *secondary*. The smaller secondary coil will develop a lower voltage than was sent to the primary coil. Since the overall power of the input and output is the same, the smaller

Fig. 11-8. Two inductors on one core.

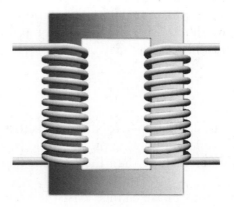

Fig. 11-9. Transformer.

secondary coil has a larger current. Likewise, when the secondary has more windings than the primary, it generates a higher voltage with less current.

This turns-to-voltage relationship is:

$$\frac{V_P}{V_S} = \frac{N_P}{N_S}$$

where N is the number of windings and P and S are the primary and secondary, respectively.

Summary

We started by looking at the magnetism generated in a coil of wire. From this electromagnet we get relays, different kinds of motors, and generators. This coil of wire can also be used by itself, as an electronic component, the inductor.

The inductor is a component that uses magnetism much like the capacitor uses the electric field. Inductors store energy in their magnetic field and release this energy when they discharge. Where the capacitor allows AC signals through but blocks DC, inductors pass DC but resist AC signals.

Inductance is a byproduct of a moving magnetic field. While the inductor component relies on *self-inductance*, you can add another coil of wire to the component and create a transformer using *mutual inductance*.

Transformers are like electromagnetic gears. By changing the ratio of primary input windings to secondary output windings you can increase or decrease the output voltage. Since the power coming out of the transformer is the same as the power coming in, the change in the output current is the inverse of the change in the output voltage.

Quiz

1. Run a current through a coil of wire and what do you get? Add a second coil and what is it then? What are a few other forms for this?
2. What do inductors store? How do they behave in an AC or DC environment?
3. Add a resistor to your inductor and what do you get?

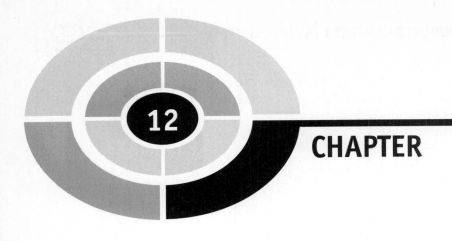

CHAPTER

12

Semiconductors

Introduction

This is the last electronics chapter, and there is a lot of territory to cover in it. At this point you should have a sense of the forces at work in electronic circuits, and some ways that these forces can be manipulated.

Most of the electronic components so far have been *passive* components. They sit in the circuit and react to the voltage and current flowing through them. Capacitors store and release energy, as do inductors. Resistors resist. Passive circuits are then circuits made up of passive devices, such as the filters explored in Chapter 10.

While passive components help us understand electronics, and are also vital to the proper functioning of circuits, the components that do must of the work in modern electronics are *active*. An active device can change its behavior dramatically based on its situation. Diodes are the simplest active device, acting like a one-way valve. The chapter explores the behavior of some active components, looking into their deepest secrets and seeing what makes them work.

Conductor Physics

Let's revisit the concepts of conductor and insulator for a moment. In Chapter 4 we stated that a conductor is an element whose electrons can move freely and an insulator is one whose electrons are stuck tightly in place.

Now we look at this electron motion just a little bit closer. Keep in mind that physics at this low level can get very tricky, so we make simplifying assumptions and gloss over a number of details.

An atom, you recall, has a nucleus of protons and neutrons. This nucleus has a positive charge. Whizzing around that nucleus is a cloud of electrons, normally just enough to make the atom's electric charge neutral.

Earlier, we hinted at the concept of *orbital shells*. The electrons are not free to orbit just anywhere, otherwise they would just spiral into the center and stick to the nucleus. Each electron has an energy level which could refer to how fast it is moving (but doesn't really), and these energy levels are almost digital in nature. An electron could have an energy of "1" or "2" but never "1.25." This discrete separation of energy levels is known as *energy quanta,* and quanta lead us into the realm of quantum physics.

Each orbital shell is associated with a particular energy level. Only electrons with the correct amount of energy can live in a given shell. If you add a quantum of energy e to an electron by, perhaps, banging on it with a hammer or shining a light on it, the electron can no longer stay in its current shell but must jump to a higher energy shell (Fig. 12-1).

Each orbital shell can hold a limited number of electrons and no more. The inner shells of an atom are normally all full. We work with the outer

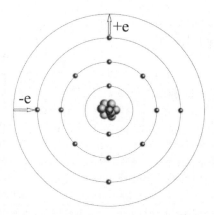

Fig. 12-1.   Silicon atom.

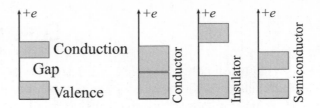

Fig. 12-2. Valence and conduction bands.

shells because they are often not full. If we add energy to an atom, an electron will jump from an outer shell to an even higher energy shell. Similarly, that electron will drop back down to its regular level when it loses its energy.

The "normal" outer shell of an atom is the *valence* and the energy level of that shell is its *valence band*. The next band out is the *conduction band*. When an electron is in the conduction band, it is very loosely bound to the atom and can be easily pushed into a neighboring atom's conduction band. The amount of energy needed to jump an electron from the valence to the conduction band is the *band gap* (Fig. 12-2).

Conductors have a very small band gap so that the energy available at room temperature kicks a bunch of the conductor's electrons into the conduction band. A solid made up of these atoms has a loose sea of electrons that is easily shifted around. An insulator has a large band gap. It takes a lot of energy to kick an electron out of its valence band and into the loosely held conduction band. Semiconductors have a medium-sized band gap and are mostly insulators.

Semiconductor Physics

Silicon (Si, element 14) is a semiconductor. At room temperature, silicon conducts a little bit. The energy available during normal operating conditions kicks some of its valence electrons into the conduction band. Not many, maybe one in a billion, but enough to create a perceptible current.

The colder the silicon is, the less it will conduct because the electrons lose their energy and fall back into the valence band. A simple component, the *thermistor*, takes advantage of this feature, changing its resistance based on the temperature.

The conduction behavior of a silicon semiconductor can be adjusted by alloying it with small amounts of impurities, or *dopants*.

DOPED SILICON

Imagine, if you will, a tetrahedral lattice of silicon where each atom is bound to four other silicon atoms. If this lattice were squashed flat and drawn in a schematic form, it might look like Fig. 12-3.

Since only the outer shell interacts with the neighboring atoms, we simplify the atom and draw only this outer shell and its four electrons. Silicon creates *covalent* bonds with four of its neighbors, meaning they share electrons in the valence shell. The valence shell in silicon has four electrons but has room for four more. It is these empty slots in the shell that are filled by electrons from neighboring atoms. This creates a tight bond between the atoms and keeps the electrons firmly in place.

What if you added some phosphorus (P, element 15) to the silicon? Phosphorus has five electrons in its valence shell, but will still form covalent bonds with silicon. Figure 12-4 shows this, though it is hard to see. With five electrons but only four of them tied in a covalent bond, there is an unbound electron. This electron is free to jump into the conduction band, giving this alloy more free electrons.

A semiconductor with these free electrons is an *n-type* semiconductor. Note that the material still has a neutral charge, the same way that copper has a neutral charge. There are simply more loose electrons available to move when a voltage differential is created.

What if you added some boron (B, element 5) to the mix? Boron has only three electrons in its valence shell, but will still form covalent bonds with silicon. Figure 12-5 shows this. In this case, there are a number of covalent bonding positions that are not filled by the boron. These *holes* are places where a free electron can get stuck. This type of alloy is a *p-type*

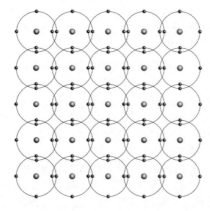

Fig. 12-3.   Silicon crystal.

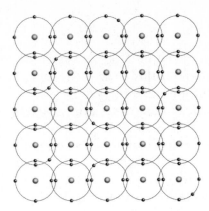

Fig. 12-4. N-type semiconductor (silicon and phosphorus).

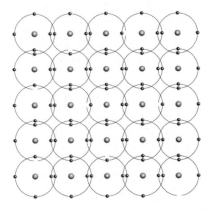

Fig. 12-5. P-type semiconductor (silicon and boron).

semiconductor. Again, it is electrically neutral. It is interesting to note that electron holes are mobile in the same way that free electrons are. As such, they are considered *charge carriers*. In most explanations, electron holes are said to carry a positive charge.

Holes and free electrons occur naturally in the silicon lattice, as well as because of the added impurities. The free electrons and holes created by the dopants outnumber these natural ones, and are known as the *majority carriers*. They carry the majority of the charge. Any natural free electrons and holes in the lattice are the *minority carriers* and provide a small background "leakage" current.

Even though Fig. 12-3 through 12-5 have been simplified down from the complex model in Fig. 12-1, it is still hard to see the differences at a glance. We don't need to see the silicon as an atom at all. We can condense the picture down to the pure electron grid, as shown in Fig. 12-6. And even this

Fig. 12-6.   Silicon electron grid.

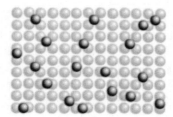

Fig. 12-7.   N-type grid.

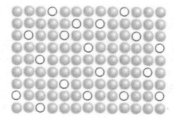

Fig. 12-8.   P-type grid.

silicon grid is just a backdrop where we can place free electrons for n-type material (Fig. 12-7) or where we can show the gaps, or holes, in p-type material (Fig. 12-8).

Diode Physics

Diodes illustrate most of the principles needed to understand the other active components. They use both n-type and p-type silicon to create a one-way valve for current. Let's look at how this works.

First you need to be thinking about the semiconductor correctly. Once you assemble a semiconductor crystal you stop thinking about it in terms of individual atoms. The crystal becomes a mass of nuclei and a sea of electrons, held together by a continuous field of force. Everything pretty much balances out. Electrons and the *lack* of electrons, the holes, can be shifted around by applying a charge to the crystal.

Most explanations treat electron holes as if they were some kind of unparticle to be moved around. What happens, though, is that the background grid of electrons shifts around and the need for electrons, the hole, ends up where the electrons aren't.

For the first step, take two tiny pieces of semiconductor, one n-type and the other p-type. The moment you bond them together they become a single crystal and electrons are free to wander between the two halves (Fig. 12-9). Note that the two halves have different properties. One has a surplus of free electrons and the other has holes where free electrons can get stuck. Remember that both the free electrons and the electron holes can move around in the crystal.

Free electrons from the N side of the crystal wander into the P side and, if they encounter an electron hole, get stuck there. Of course, as the electrons on the P side drift away from the junction, the holes appear to wander toward it and into the N side, gathering more free electrons. Every time a free electron from the N side gets stuck in the P side, the N side loses a bit of negative charge and the P side gains a bit. This process continues until the forces balance and there is a strip in the n-type semiconductor where the free electrons have been removed, leaving it with a positive charge. Those errant electrons are all trapped in holes in the p-type semiconductor, giving it a strip of negative charge (Fig. 12-10).

Overall the material is still neutral. In fact, the electrons trapped in the holes are keeping the other electrons away. They repulse the free electrons in the n-type side so they stay away from the barrier. The positive strip is

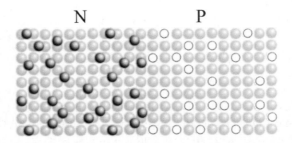

Fig. 12-9.   N-P junction (first joined).

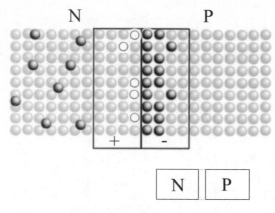

Fig. 12-10.   N-P junction (depletion region).

an illusion, since it is just the lack of electrons. The barrier electrons also repulse those in the semiconductor lattice in the p-type side, which is why the "holes" flow toward the barrier. Eventually, though, the forces all balance out.

The barrier of locked electrons is the *depletion zone*. The free electrons have been depleted, and the holes have been taken away (filled) so they can be considered to be depleted too. While the two halves of the depletion zone may have electrical charge, it is not a movable charge. Everything is stuck in place, so the depletion zone is an insulator.

FORWARD BIAS

The next step in creating a diode is to attach two wires to it, one on each side. When you apply a voltage across the diode in the forward direction, the current flows. Why?

Figure 12-11 shows the forward bias circuit. The battery, or other power source, is pushing electrons into the N side of the diode. As external electrons are pushed into the N side the pressure builds up until it begins to overcome the repulsion of the depletion zone. This shrinks the depletion zone and electrons are free to move again.

The electrons continue to move across the barrier. They can kick electrons out of their holes and out the other side of the diode. This gives us current through the diode. The pressure needed to shrink the depletion zone takes away some of the pressure from the circuit. This is the voltage drop across the diode.

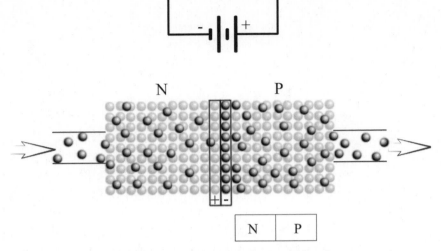

Fig. 12-11.   N-P junction (forward biased).

REVERSE BIAS

When you hook up the battery backwards (Fig. 12-12) the situation is reversed. The applied charge makes the depletion zone larger and even harder to cross. Pushing electrons into the P side of the diode provides more free electrons in the P side. This increases the negative charge at the P side of the P-N junction, which expands the positive charge on the N side.

Some electrons still make it through, but they are fighting the binding force in the silicon's valence so not many make it. This is the leak current you get when a diode is hooked up backwards.

Electronic Switches

ANALOG VERSUS DIGITAL

In Chapter 7 we made a distinction between analog and digital signals. Everything we have looked at so far has been analog in nature. This chapter introduces some digital concepts.

A digital signal represents the state *on* and *off*, 1 and 0. Whenever a voltage is above a defined value it is on. Whenever it is below a different, lower, value it is considered to be off.

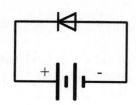

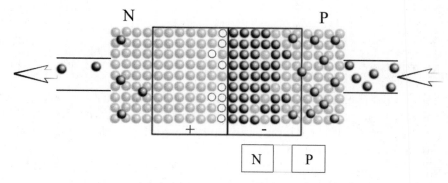

Fig. 12-12.   N-P junction (reverse biased).

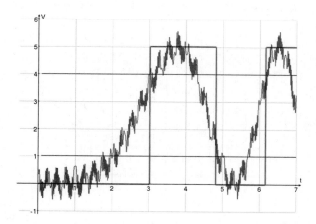

Fig. 12-13.   Analog and digital signal.

Traditional digital electronics operate between 0 V and 5 V, though many modern systems work at much lower voltages. In the 5 V case, *on* may defined as being a voltage greater than 4 V and *off* may be any voltage below 1 V (Fig. 12-13). Voltages between these high and low *thresholds* of 1 V and 4 V are undefined.

Note that this is only an example. Each component specifies its own minimum, maximum, and threshold voltages. The switching behavior as the voltage changes from high to low also varies based on the component in question.

We've already looked at one kind of electronic switch, the relay. Relays are what are known as *electromechanical* devices. They use electrical energy to drive a mechanical mechanism, in this case, a switch. A true electronic switch has no moving parts. One such switch is the transistor. Another is the MOSFET.

TRANSISTOR

A *transistor* behaves like a variable resistor or, using the water analogy, a water faucet. The basic transistor is the bipolar junction transistor, or BJT. Transistors come in two flavors, npn and pnp. Note that transistors are not digital, but are analog switches.

As the name suggests, an npn transistor consists of three pieces of material, a thin p-type semiconductor with n-type semiconductors on either side (Fig. 12-14). This is like two diodes placed together back to back (Fig. 12-15). The transistor's schematic symbol even reflects this two-diode shape. Using

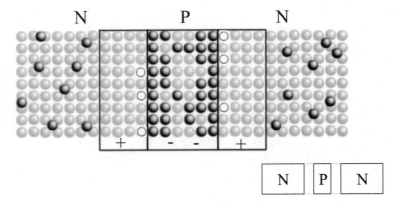

Fig. 12-14. NPN transistor.

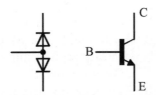

Fig. 12-15. NPN transistor symbol (two diodes).

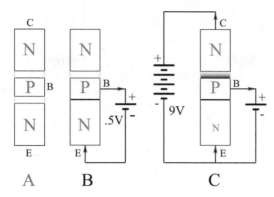

Fig. 12-16.   NPN transistor operation.

just the block diagrams that show the depletion zone, let's do a quick tour through the transistor's operation. The sequence is shown in Fig. 12-16. Fig. 12-16A shows the raw N-P-N configuration. Note the depletion zones on both sides of the P section.

In transistor terminology, the center of the transistor is the *base* and is marked with a *B*. One of the ends is the *emitter* (*E*) and the other is the *collector* (*C*). If you connect the emitter and collector into a circuit right now, no current would pass through the transistor because the depletion zones insulate these sides from each other.

What if you were to apply a small, say half-volt, charge between the base and the emitter (Fig. 12-16B)? These two sections, by themselves, are a diode and you would have forward biased it. The insulating layer between the emitter and base is reduced just enough so that it can conduct.

Now, what if you connect a larger charge, say 9 V, bewteen the emitter and collector (Fig. 12-16C)? The emitter-base diode is not directly affected, since we aren't letting more than a half volt of charge out of the base. The collector develops a large positive charge as its electrons escape, and there is plenty of negative charge at the emitter from the large battery.

There is, however, still that insulating layer between the base and the collector. Doesn't that stop the current? Interestingly enough, no. The base layer is thin enough, and the charge difference between the base and collector is large enough, that most of the electrons zoom through the emitter and jump across the gap into the collector. Current flows from the emitter to the collector as long as the voltage differential across the emitter-base junction keeps that insulating layer thin. If you remove the base voltage, the emitter-collector current stops. If you increase the base voltage, the emitter-base insulator is even thinner and more electrons make it across to the collector. The emitter-collector current increases.

The name *transistor* is short for *transfer resistor*. The voltage at the base of the transistor controls the current flow across the rest of it. It is, in fact, a voltage-controlled variable insulator. As fundamental as transistors are, I don't recommend using them unless you are interested in designing low-level circuits. For most electronic switch applications, the MOSFET is a better device. If you are looking to amplify a signal, you are probably better off buying a commercially designed amplifier on an integrated circuit. Both MOSFETs and integrated circuits are discussed below.

PNP transistor

The pnp transistor is the same as the npn transistor, but all of the roles are reversed (Fig. 12-17). The base is made up of n-type semiconductor and the emitter and collector are both p-type. The battery hookups are likewise reversed. See if you can work out the operation from first principles.

FET

A different approach to the transistor is the *field effect transistor* (FET). The FET comes in both npn and pnp flavors, like the BJT. Let's look at just the npn FET (Fig. 12-18).

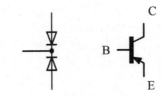

Fig. 12-17.   PNP transistor.

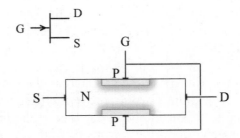

Fig. 12-18.   Field-effect transistor.

In this transistor, the two ends are the *source* (*S*), which is where electrons come in from ground, and the *drain* (*D*) where the electrons go out. The controlling area is the *gate* (*G*). With no charge at the gate, there is a small depletion layer between the N and P semiconductors. Current is still free to flow from the source to the drain.

If you apply a negative charge to the gate this depletion layer grows, pinching off the flow of electrons. The more gate charge, the more pinched, until the resistance between the source and drain is large enough that no effective current remains.

MOSFET

A MOSFET is a variation on the FET. The "MOS" stands for *metal oxide semiconductor*. The metal oxide is a glass-like layer that is an excellent insulator. It is used to completely insulate the gate from the rest of the transistor (Fig. 12-19).

In this example we look at an enhancement mode MOSFET. The source and drain are both n-type semiconductors embedded in a p-type substrate. They are isolated from each other, so no current can flow.

The gate is a metal plate separated from the substrate by a thin oxide insulator. When the gate is made positive relative to the source, the holes in the p-type substrate are pushed away from the gate (meaning the lattice electrons are attracted to it). This opens up a conducting channel, allowing current to flow from the source to the drain.

A depletion mode MOSFET is like the FET, where the source and drain are attached to a common n-type channel. The electric field at the gate pinches the channel closed. The gate of a MOSFET is almost perfectly insulated from the conduction channel, so there is no noticeable current leakage. The gate acts like a capacitor.

Where the other transistors are analog switches, amplifiers even, the MOSFET has a limited analog range. It is more useful as a digital switch,

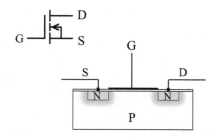

Fig. 12-19.   MOSFET.

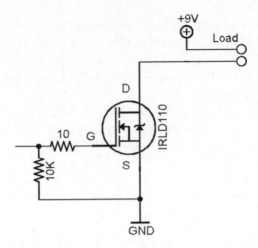

Fig. 12-20.   MOSFET switch.

turning current flow on and off. An example of a simple MOSFET switch is given in Fig. 12-20. The control signal is passed through a 10 Ω resistor whose sole purpose is to protect the MOSFET from static discharge, since the gate is easily damaged. Depending on your circuit, it may not be necessary. The 10 kΩ resistor to ground is there to make sure the gate returns to ground when there is no input signal driving it.

The specific MOSFET shown here is capable of turning on with a gate voltage of just 2 V, though the device is operating at its best between 4 and 5 V. The switch itself is across the source and drain. It can switch a maximum of 100 V at 1 amp, which is good for a $0.50 part.

Integrated Circuits

We've spent all of this chapter looking at the smallest details of semiconductor function. The transistors, while critical to all modern electronic circuits, are still just one small component. Together with capacitors, resistors, diodes, and so forth, they can be assembled into circuits that do something.

It is interesting to note that almost all electronic components can be constructed from thin layers of doped semiconductors and insulating oxides. These layers can be created fairly easily, with methods not entirely unlike developing photographs. This means that we can take a single slab of silicon and turn it into a bunch of different electronic components, all microscopic in size. And these components, still using various photographic masking

techniques, can be connected with films of metal. Wires. An entire electronic circuit can be constructed at microscopic scale. And this is an *integrated circuit (IC)*.

One last component makes our list complete, and that is the *crystal*. This is a slab of crystal that *resonates*, or prefers to vibrate, at a particular frequency. This frequency depends on the crystal's composition and size. Crystals are used to create AC signals that are then used as a heartbeat for digital circuits.

There are many different types and families of integrated circuit. Analog ICs can create amplifier and filter circuits, among others. Digital ICs perform all of the various logical operations. Some ICs convert between analog and digital signals. Large-scale ICs can put hundreds, thousands, even millions of components in one chip. The central processing unit in your computer is one such chip.

Summary

To do the subject matter in this chapter justice would take half a book. However, the details you learned in this chapter provide the foundation for all of the variations that book would explore.

We started this chapter by looking at the odd behavior of semiconductors and some of the reasons behind it. Even more interesting than simple doped silicon is how it reacts when n-type and p-type sections are joined. We then explored the behavior of the n-p junction inside a diode, the simplest active semiconductor component.

Expanding on the diode, we looked at the basic transistor and some variations. Transistors act like small amplifiers or switches, and are the foundation of all active electronic circuits.

Finally, we introduced the idea of the integrated circuit. If you want to explore further, perhaps to see how transistors are used or what types of semiconductors can be found, any book on electronics will be happy to take you further.

Quiz

1. What are the two types of electronic component? Into which category do semiconductors fall?
2. Do you remember all of the jargon relating to semiconductor physics?

3. What are the types of doped semiconductor? What makes them special?

4. What is a depletion zone? What does it do? What happens to it when it is forward biased? Reverse biased?

5. What does a transistor do?

6. Draw a picture of a field-effect transistor and remind yourself how it works. Hand-waving is optional.

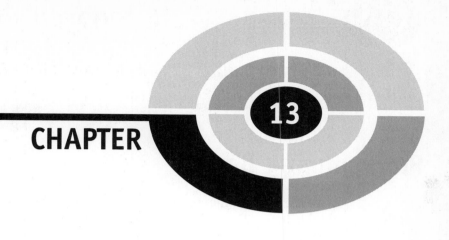

CHAPTER 13

Programming

Introduction

We've already introduced many of the concepts used in a robot's brain. Chapter 6 talked about feedback and Chapter 7 introduced the idea of sequences of actions and decisions based on outside input.

Did you miss that part? It was where a motor turned on when a switch was pressed. In technical lingo, "IF this switch is pressed THEN turn on the motor." Sure, it happened automatically due to the nature of the switch, the current flowing through it, and the motor in the circuit. But it was still a decision.

This chapter talks about machine behavior as a computer program. It explores sequences of actions, controlled by software instead of hardware, and it works with timing and feedback. The programs are written in the simplest environment available, using the programming graphs from LEGO

Mindstorms. So charge up your batteries and read on for some hot robot action.

Programming Basics

There are many different programming languages, each with its own syntax and semantics. *Syntax* consists of the words, symbols, and basic grammar of a language—the computer's instructions and how they are put together. *Semantics* is the meaning behind the instructions, how they behave when they are executed. A program is a set of instructions that tell the controller, or computer, to do something.

Most programs are written using just a few basic concepts, though as in all things there are exceptions. The programming style we explore in this book is known as *control-flow*. In control-flow programming the computer's attention shifts from one instruction to the next, flowing through the program. The focus of the program is on the flow of commands.

An alternative model is *data-flow*, which behaves like an electronic circuit where the electric current is information and the components in the circuit are program instructions that modify the data. The focus of the program is on the flow of data.

FLOWCHARTS

Programs have a "shape" that illustrates the flow of control. The simplest way to see these shapes is with a *flowchart*, which is a schematic of your program. There are many different ways to diagram a computer program, and the flowchart is one of the original techniques. It remains useful for describing the behavior of simple programs.

Flowcharting is part of the *Uniform Modelling Language* (UML). Data-flow diagrams are too, as well as timing diagrams, object hierarchy and colloboration diagrams, and others that we don't need to be concerned with now. An example flowchart is shown in Fig. 13-1.

$$\boxed{\text{Terminal}}$$

Each program has a beginning and most programs have an end. These are the *terminal* states of the program and are labeled, sensibly enough, as

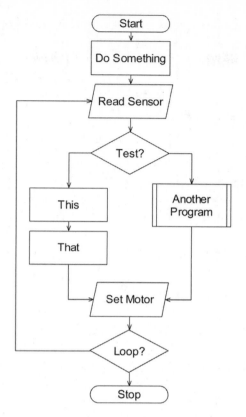

Fig. 13-1. Flowchart.

"start" and "stop." The start terminal could also be named something that describes the program, such as "Motor Control."

From the start terminal, control passes through a *flow* arrow and to the next box. The arrows always point to the next box to execute in the flowchart. This represents the flow of control through the program.

Process

Most of the instructions inside a program do something. They read or write data in memory, perform some math, and otherwise process information.

Process blocks contain one or more instructions that do something to achieve a result. The label of a process block should describe that result, such as "Initialize Motor Control."

Sometimes your program needs to access information outside of its own memory. These accesses are performed by instructions in the program, and as such could be shown as process blocks. An outside reference, however, is a significant event so these *input* and *output* (I/O) blocks have their own shape.

Decision

A program that runs straight through from start to stop, executing one or more process blocks in between, is not a very interesting program. More often than not, you want to make decisions and perform different actions based on the result.

These decisions are usually formed as a yes/no test, such as "Is this value is greater than 10?" If it is, the "yes" flow is followed, othewise the "no" flow is followed. Another way to state this is as an *if* statement, "*If* this value is greater than 10 *then* do the yes actions *else* do the no actions." This is also known as a *branch* in the program, since there is one flow coming in (the trunk) and two flows (the branches) coming out.

If one of the branches points back up the flowchart, against the normal top-to-bottom flow, it is called a *loop*. Loops let you run the same section of code over and over again forever, or until some condition is fulfilled.

Sub-Program

Sometimes you want to do a lot of work in one branch of a test, so much work that showing it all would overwhelm the flowchart. It is best to keep any given diagram the size of a single page, so it is easy to understand. When you can't keep it simple, you can break out a chunk of the program into its own *sub-program*. This would have its own flowchart that can be included by a reference in a higher-level chart. Another name for a sub-program is *subroutine*.

RCX PROGRAMMING

The environment we are programming in is the LEGO Mindstorms Robotics Invention System 2.0. The heart, or should I say *brain*, of the system is the RCX controller. Programming the RCX is a lot like drawing a flowchart. This makes it easy to snap together programs.

Try This: An RCX program that mirrors the flowchart in Fig. 13-1 is shown in Fig. 13-2. The first thing this program does is turn on motors A and C.

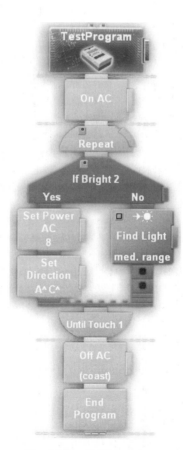

Fig. 13-2. RCX program.

The *Repeat...Until* is the RCX representation of a loop. In this case, it loops until the switch on input 1 is touched.

Inside the loop is an *if-then* block, known as a *Yes or No* block. If the light sensor on input 2 is bright, the motors are set to full power forward, otherwise we execute a complex subroutine to find the light. Once the loop is completed, meaning the robot has bumped into something, the motors are turned off and the program is stopped.

A version of this program that is easier to read in this black-and-white book format is given in Fig. 13-3. We shall use this schematic form through the rest of our RCX programs. The processes in this flowchart are RCX *Small Blocks*. *Big Blocks* or *MyBlocks* are sub-programs. *Wait* and *Repeat* structures are shown with a new schematic shape so you can distinguish them from tests and processes.

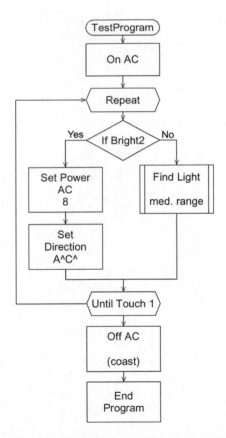

Fig. 13-3. RCX program, flowchart style.

Programs

SIMPLE TIMED SEQUENCE

Try This: The simplest program is a sequence of timed actions. *Do this for a second, then do that, then do this other*, and so on. One example of this is the *Dance* program illustrated in Fig. 13-4. This program is controlling two motors named A and C. They can be set to move forward "ˆ" or backward "v" at different power levels. They could be in any robot with *differential drive* steering, where the speed and direction of the drive motors are used to steer the robot. The program is not complex, though it does create a fancy little dance routine. It's long, so long that it had to be broken in half to fit it on the page, but it's simple.

"Dance" uses timed behaviors. *Go forward for a second, then go backward for a half second, then turn for a half second*. While timed behavior is useful, it is still open-loop control. The distance traveled in a half second may be different depending on the state of the batteries, the weight of the robot, what kind of floor it is on, and perhaps even the phase of the moon. There is one loop in this program. It repeats the entire string of behavior once, meaning it runs it two times, the first pass through and the one repeat.

A more interesting program might use feedback to keep the motions consistent. If you had a rotational sensor, not included in the standard kit, you could count how many times each wheel turns. This count can be used to decide when the robot should change directions. Even simple feedback like counting wheel rotations can make your robot more predictable.

OBSTACLE AVOIDANCE

Try This: One of the simplest feedback-based programs is the basic bumper-car program. This program assumes your differential-drive robot has two bumpers on it, sensors 1 and 2. This version of the program drives forward until it hits something. Once it has hit an obstacle, it backs up and turns away from it and then resumes driving forward. The first part of the program is shown in Fig. 13-5.

The first blocks set some variables. A *variable* is a named area in memory that can hold a value. When the variable name is used, it represents the value in its associated memory. That value in memory can be changed at any time, so the next use of the variable uses the new value. A named value that never changes is called a *constant*, but we don't use those here.

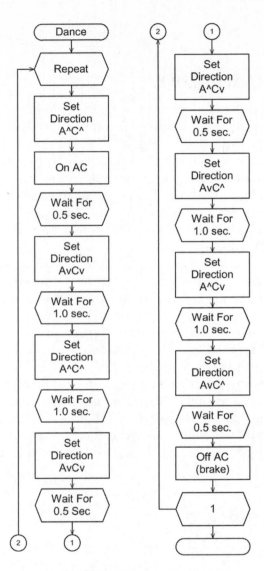

Fig. 13-4. Dance program.

Instead of setting and using variables, you could sprinkle the actual numbers through the program. There are benefits to using variables instead of numbers. First, it makes the program read better. "Set Power AC 4.0" isn't quite as clear as "Set Power AC PowerHigh."

More importantly, variables make the program easier to modify. What if you want to have high power be 8 instead of 5? Change the variable at the

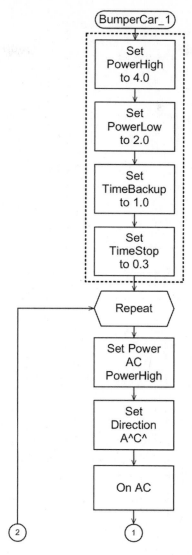

Fig. 13-5. BumperCar program 1 (part 1).

top of the program. Without the variable, you would have to find each instance of the power value 5 and change them all.

Note the dotted lines around the first four program blocks. These blocks are all part of a single conceptual action, *initialize variables*. We could collect these into a subroutine called *Initialize* and make the flowchart smaller without losing much of its meaning.

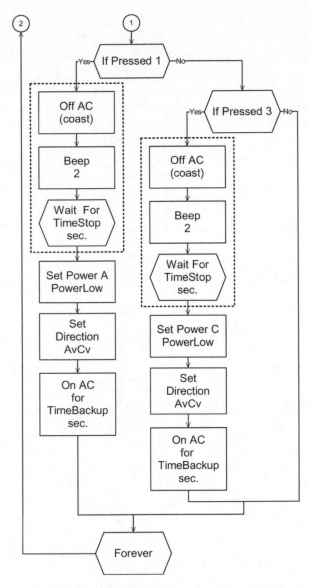

Fig. 13-6.   BumperCar program 1 (part 2).

Once the variables are set we drop into a *Repeat* loop. The first action in this loop is to set the robot to move forward at high power.

Next, we make decisions based on input from the outside world (Fig. 13-6). The first checks if switch 1 has been pressed. If it has, then it stops the robot and beeps. After a brief pause it backs up in a curve. If button 1 was not pushed, then we check button 3. If that button is pushed we repeat the stop-and-backup sequence of actions, but along a different curve. This test-inside-a-loop logic is known as *polling*. The program looks at the switches whenever it can, during the normal course of the program.

Note the two sets of processes surround by dotted lines. These both contain exactly the same sequence of steps. Repeated code like this is a good candidate to be replaced by subroutine. If you decide to change the behavior of the timed stop, you can edit the code in one place, the subroutine, and the effect will be seen everywhere it is used. Once the decision-making and backing-up code is processed, we return to the top of the loop, repeating forever.

A compressed, easily readable version of this program is given in Fig. 13-7. Packaging-related blocks of code makes the program more readable, and it helps you to reuse common blocks of code. A side effect is that you can't see all of the low-level program details without opening up the subroutines.

Parallel processing

While most computers and controllers can only follow one flow or *thread* of code at a time, a fast controller can pretend to follow more than one simultaneously, or in *parallel*. A program with more than one thread executing at one time is *multi-threaded*.

We can organize the program in Fig. 13-7 into three simpler programs. Note that every programming environment has its own rules for parallel, or multi-threaded, programming. The RCX rules are very simplistic and involve the use of sensor watches. Only the watching part is done in parallel.

Figure 13-8 shows our restructured program. Note that I am using the flowchart I/O symbol for the watcher blocks. When the program first starts, it creates two "watcher" threads. The first one continuously tests to see if the first switch is pressed, and the second thread watches switch 3.

With the watchers firmly in place, the main program runs and ends up in an endless forward-moving loop. In the background, the watcher threads are waiting for a switch to be pressed. When a switch is finally pressed, the watcher takes control and runs its own program. When the watcher thread is done it returns control back to the main program.

This is a form of *interrupt* programming, where an event can interrupt one program in order to run a different sub-program. When the interrupt has been processed, it then returns control.

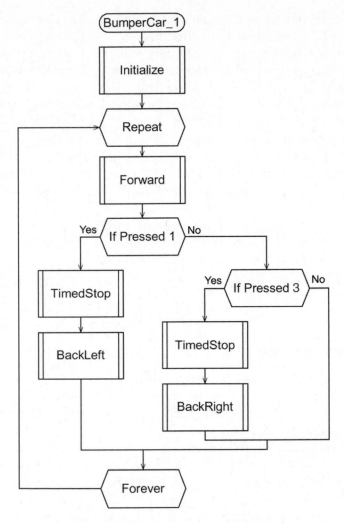

Fig. 13-7.   BumperCar program with MyBlocks.

LINE FOLLOWING

Try This: A similar feedback system can be found in the traditional line-following robot. What is a line except an "obstacle" that has been painted on the ground? Instead of bumper switches, the robot uses a light sensor to "see" the line.

The basic line-following program, found in the Mindstorm examples, "bounces" the robot off the line (Fig. 13-9). Our only change is to add a *Set Power* block at the top, so you can adjust the robot's speed.

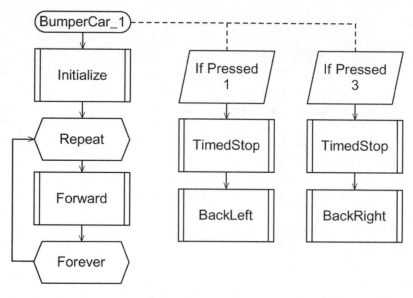

Fig. 13-8.   BumperCar with watchers.

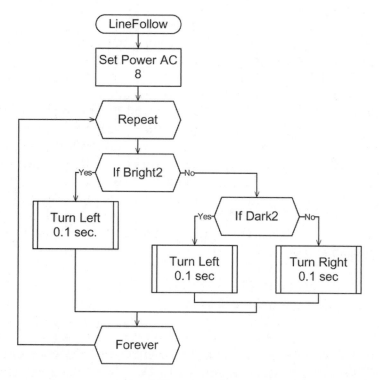

Fig. 13-9.   Line follower.

There are several hidden lessons in this example. The first one is that it is easier to start with existing code than to begin inventing everything from scratch. It is easy to get excited by technology and run off to create everything from scratch, just so it is "all yours." Resist this impulse! Look around and see what other people have done, and it will save you hours, days, or even weeks of reinventing the wheel. Once you've found a good first approximation for your solution, you can take all of that extra time and adjust the found solution to work better. In the end, you'll have a better solution for the same amount of time invested.

Another lesson is that a good solution to the problem may not be the obvious solution. Note that the robot turns left if it is on the white background and it turns left when it sees the line, too. It only turns right when it is on the edge, between bright and dark. There are three states in the sensor, as programmed by the LEGO people: Bright, Dark, and somewhere in between.

Why does this program work? What happens if you remove the *If Dark 2* test and just turn right when it's not bright? How does it behave if you have a fast robot? A slow one? This program doesn't follow the line efficiently, since it spends more time turning than it does moving forward. What happens if, instead of stopping the inside wheel, it sets it to move at a slower speed?

Moving too fast can be a problem since you can overshoot the line and its corners. In all robotic applications, there is a middle ground where the speed of the machine, the response time of the sensors, and even the momentum of the robot come together to make a working solution.

It often takes a lot of experimentation to find just the right combination. This is where the scientific method comes into play. First, do some tests with an existing system to see how it behaves. Then develop a theory on how to make it better. Test this theory and see if it works. If it does, great! Otherwise, make a note of why it didn't work... and try again with a different theory.

When you are making changes in your theories and tests, take care to change just one or maybe two things at a time, so you can tell which change affected the result. If you try a bunch of stuff all at once, you won't know what each change did.

Two sensors

If you have two light sensors, you can make your program faster and smarter. These sensors would act like two bumpers. As the left sensor crosses the black line, you can turn the robot left. For the right sensor, you would turn right.

Light following

These sensors don't have to be pointed down to the floor. You can point them forward or upward and use them to follow the light from a flashlight.

SELF-CALIBRATION

Though the RCX software handles light sensor calibration for you, there are times when you want to do this yourself.

Most sensors do not return information in useful forms like "bright" and "dark." Instead, they return values in a numerical range, such as 0 through 255, or 0.0 to 1.0. The values depend not only on what is in front of, or touching, the sensor but also on the ambient environment. If a value of 40 is a bright reflection in a normal room, what would it be if the robot is in a brightly lit room? If we are thinking about the reflective light sensor, what is a bright reflection on paper? The floor of your kitchen?

To turn the arbitrary, and locally varied, values returned from a sensor into valuable data you need to *calibrate* the sensor. For a reflective light sensor, you need to know what values it returns for the bright surface as well as a dark line. A simple way to do this is to look, reading the light sensor as the robot spins in a circle. Remember the largest and the smallest values from the sensor for your definition of bright and dark (Fig. 13-10).

You don't want to test against these mininimum (min) and maximum (max) values directly. You want to see how close the current light value is to one of them, or the current value's relative position from the midpoint. Once you have captured your light values, you can stop the motors for a moment to let the robot come to a stop, and then turn back to the starting position.

If you want to make your program really clever, though more complicated, you can keep track of the light values as the program performs its normal duties. You can then use these values to keep the min and max ranges up to date as conditions change over time.

Summary

Programs consist of a series of commands to a controller. These commands can perform some action, manipulate memory, perform math, sense the environment, or make decisions based on internal or external conditions.

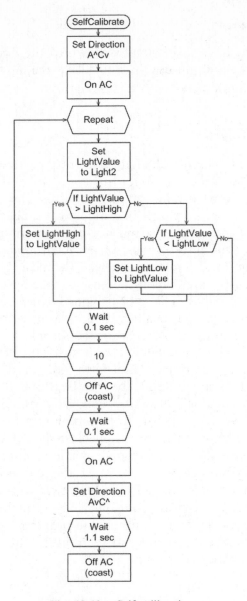

Fig. 13-10. Self-calibration.

From these basic building blocks you can create any manner of interesting and complex behaviors for your robot.

Feedback lets your program, and hence robot, react to the environment around it. Whether it is responding to bumper switches or varying values from a light sensor, your program can incorporate feedback to keep it moving in the right direction.

Of course, most robots are not little wandering machines. Regardless of the job a robot does, or the shape it takes, it will be controlled by a program and it can use feedback from sensors to keep itself running smoothly.

Quiz

1. What are syntax and semantics as they relate to computer programming?
2. What are subroutines and what are they good for?
3. If your program keeps watching an input waiting for it to change, what is that called? What if the input interrupts the program, forcing it to pay attention?

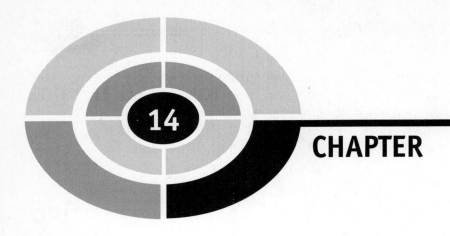

CHAPTER 14

Shaping Motion

Introduction

In Chapter 6 we talked about passive control. In this chapter we explore some of the mechanisms that can take simple rotation and turn it into complex motions.

This chapter makes use of linkages—bars or rods that are linked together—to control the shape of the motion. You can use clever arrangements of links and gears to make the best use of your power sources. This chapter explores some of those clever arrangements, starting with simple mechanisms and moving into more complex ones.

Given a choice between using one motor and a complex mechanism, or two motors and perhaps a simple mechanism, the one-motor solution will probably cost less and be more reliable.

Looking Back

We've already discussed a number of mechanisms that shape motion. The cam leaps to mind, since the varying distance of the outside edge to the center axis is used to push a cam follower in an arbitrary way.

The simple gear or pulley train shapes motion if you consider how it slows down or speeds up rotation.

Sprockets and chains do the same thing as gears with the added bonus of converting the rotary motion of the sprocket into the linear motion of the chain. This linear motion can be used to move things from one place to another, like a conveyor belt or escalator, or perform some other activity.

The transmissions shaped motion, as did the levers and U-joints... it's what machines do. Here we explore some more ways to shape motion.

Single-Link Mechanisms

Try This: A single-link mechanism can be something as simple as a lever (Fig. 14-1). A lever can be used to reverse the direction of motion, so when you push down it pushes up. A lever can also transform the length and strength of the motion.

A rotating link is more like a cam than a lever (Fig. 14-2). It can be used to press on another link, creating an intermittent back and forth movement, or *reciprocal motion*.

If you stick a second link on this rotating lever, you are moving into the world of two-link mechanisms. However, if you look at it like Fig. 14-3 it might still seem like a one-link mechanism, with one link attached to a wheel. In this use, the wheel and the rotating link are the same thing. The mechanism in Fig. 14-3 converts rotating motion into a continuous reciprocal motion.

Fig. 14-1. Single link (lever).

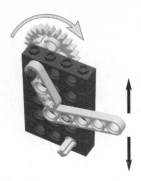

Fig. 14-2.   Cam and lever for intermittent motion.

Fig. 14-3.   Rotary to reciprocal motion.

A variation on the lever from Fig. 14-1 is the bent lever. This transforms motion by 90° instead of 180° (Fig. 14-4). This is a known as a *bell crank* since it was used in the bell-ringing systems of times past. The angle between the two arms can be adjusted to change the angle of output relative to the input. The extra links shown in this figure are there to get force into and out of the bell crank. These could be strings, pushrods, or anything else and don't count as a "link" in our accounting.

A variation of Fig. 14-2 provides continuous reciprocal motion in a lever using a different approach (Fig. 14-5). In this case the cam lifts and lowers the lever from below instead of tapping it from above.

You can also use a lever pinching against a fixed structure to provide some dynamic gripping action (Fig. 14-6). In this gripper, the bottom beam is fixed

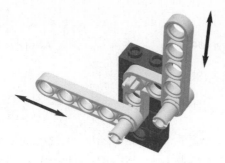

Fig. 14-4.   Motion redirection.

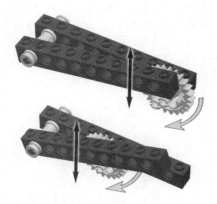

Fig. 14-5.   Cam and levers for continuous motion.

Fig. 14-6.   Single-link gripper.

rigidly to the framework and the top lever can rotate up and down to open and close the "claw."

Two Links

Try This: With two links, your motion options become more interesting and better behaved. For example, the reciprocal motion from Fig. 14-3 can be made more *linear*, or line-link, using two links (Fig. 14-7).

This two-link mechanism should look familiar, since it is found on steam locomotives. In that case, the linear motion of the steam-driven piston pushes on the linear link which then turns the drive wheel of the train. There are some difficulties in converting linear motion to rotation, which you will discover as soon as you try it yourself. The fix is to attach two cylinders to the wheel, with the second one attached 90° from the first. When one link is in the "dead" zone, the other is ready for its push.

Fig. 14-7. Better rotary to reciprocal motion.

Fig. 14-8.   Two-link synchronized gripper.

The gripper from Fig. 14-6 can have a second moving link. The trick is to keep the two links moving together without having to add much complexity to the mechanism. This is done by putting gears at the base of the links as shown in Fig. 14-8. While the two-link gripper is an obvious extension of the one-link gripper, two links can be assembled to create novel motions as well.

The two link system in Fig. 14-9 creates a "D"-shaped pushing motion that can be used for any number of applications. The bent link directs the motion to where it will be useful. The shape of the motion depends on the relative position of the pivot points and the lengths of the two links.

Given a desired motion, it should be possible to calculate the desired link sizes and positions, but that is a nontrivial task. We will save our difficult calculations for the more generally useful four-bar linkage in the next section.

More Links

PARALLEL MOTION

Try This: The four-bar link provides a way to move a link parallel to a base (Fig. 14-10). In this accounting, the base is considered a bar—creating a flexible box.

Fig. 14-9. Two-link pusher.

Fig. 14-10. Four-bar parallel motion.

Two of these parallel links, plus the gearing system from Fig. 14-8, can make a parallel-gripping hand (Fig. 14-11).

The parallel mechanism can be turned in the other direction for a narrower grasp (Fig. 14-12).

A gearing arrangement can replace the bars to provide parallel motion, though you then need to contend with backlash. Figure 14-13 shows geared parallel motion. The arm is driven from the side by an additional link.

Fig. 14-11.   Parallel gripper.

Fig. 14-12.   Another parallel gripper.

Fig. 14-13. Geared parallel motion.

FOUR-BAR LINKAGE

The four-bar linkage can be used for more than simple parallel motion. By changing the length of the four links you can "program" the linkage to start in one position and end in another.

First we need to name the links for easy reference. The base or stationary bar is the *ground bar*. The link whose motion we are designing is the *coupler bar*. The other two links are the *crank* and the *follower*. A crank is the powered link, and a follower is the unpowered link.

The first thing to do is to decide where the coupler bar must go. Only the starting and ending positions are significant, so we draw these into place relative to the ground bar (Fig. 14-14).

Let's call one end of the coupler A and the other end B. Draw a large circle centered on the start and end positions of end A. The circles need to be large enough so that they overlap, as shown in Fig. 14-15. Now draw a line through the two points where the circles overlap. Where this line crosses the ground bar is where the follower is attached. The other end of the follower is, of course, attached at point A on the coupler. The length of the follower is the distance from the ground pivot to point A. This procedure is repeated with point B (Fig. 14-16).

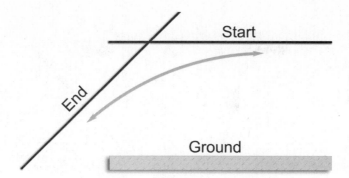

Fig. 14-14.   Desired start and end position.

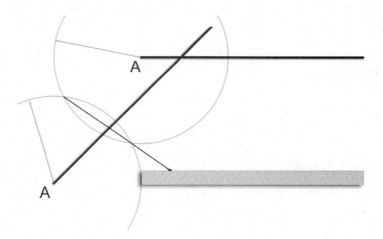

Fig. 14-15.   Locating the follower pivot.

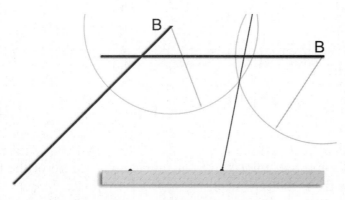

Fig. 14-16.   Locating the crank pivot.

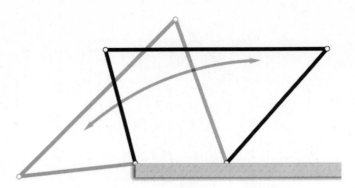

Fig. 14-17. Complete four-bar linkage.

When you build this mechanism, it will reach the desired positions precisely, as shown in Fig. 14-17.

COMPLEX MOTIONS

Try This: The oscillator in Fig. 14-7 provides a smooth, reciprocal motion that looks something like Fig. 14-18.

Coupling two cranks together to drive a follower link can provide a more complex, *harmonic motion*. One such mechanism is shown in Fig. 14-19. An estimate of its motion profile is given in Fig. 14-20.

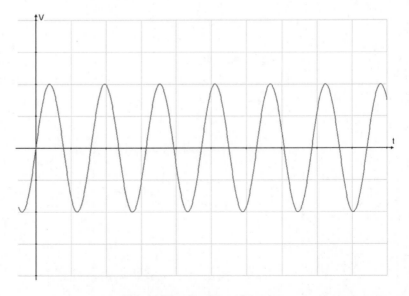

Fig. 14-18. Reciprocal motion.

Fig. 14-19.   Harmonic linkage.

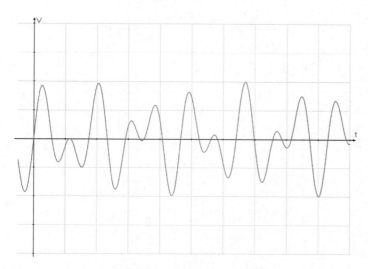

Fig. 14-20.   Harmonic motion.

Other Mechanisms

If you spend some time browsing through online patents, either via the Patent Office (www.uspto.gov) or some other patent portal like Delphion (www.delphion.com, and yes they have a free version buried in there), you will see numerous clever and unique mechanisms to achieve many different mechanical goals. Some examples are illustrated here.

CARDAN GEAR

While the the crank-and-piston arrangment can create linear motion from rotation, the *cardan gear* does this in less space and with a completely different technique (Fig. 14-21).

   The cardan gear is actually two gears. The outside ring is a gear turned inside out and it provides a fixed framework for the moving gear. The inside gear has exactly half as many teeth as the outside gear. A post positioned over one tooth of this inside gear follows a straight line. This post can be used to drive additional machinery.

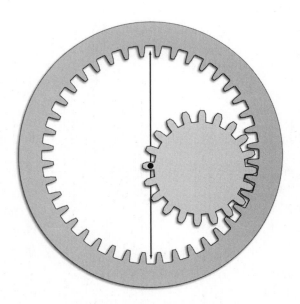

Fig. 14-21.   Cardan gear.

Fig. 14-22. Quick-return.

QUICK-RETURN

Try This: A gear with a post on it can be used to drive other mechanisms; for example, in a *quick-return* (Fig. 14-22), the gear pushes the link back and forth. At the top of its arc the post moves the link slowly, along a large radius. At the bottom of its arc the post moves the link quickly since it is closer to the link's axis of rotation.

GENEVA STOP

The *Geneva stop* gives intermittent motion (Fig. 14-23). In position 1 the Geneva gear is locked against the hub of the crank. The crank is a stud on a disk in this example. In position 1, the crank is entering the slot in the gear. In positions 2 and 3, the crank drives the gear through a partial rotation. The gear has room to move because of the crescent cut out of the crank's hub. Position 4 shows the gear locked into place again.

RATCHET

Try This: Another useful mechanism, the *ratchet*, allows motion in one direction but not the other (Fig. 14-24). It consists of a toothed wheel, like a gear, and a *pawl* that slips between the teeth. As the ratchet wheel rotates

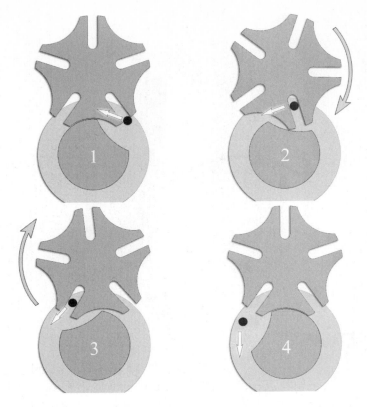

Fig. 14-23. Geneva stop.

forward the pawl slides up the tooth and, as the tooth passes, it clicks down again.

Moving backward, the tooth hits the pawl and prevents further motion. Note that the ratchet wheel is not normally a gear but will have teeth specially designed for easy motion and solid braking.

The pawl needs to be pushed down so that it catches solidly in the ratchet. Gravity can serve this purpose in some cases, or the pawl can be held down with a spring or, in the case of Fig. 14-24, a rubber band.

Summary

This chapter provided a brief pictorial view of some of the simpler mechanisms you can use to shape and control motion. We looked at mechanisms

Fig. 14-24.   Ratchet.

from the simple lever to a geared link system providing complex harmonic motion.

A large number of the mechanisms acted to change the direction of motion, either from a linear motion to a different linear motion, or from rotary motion to a linear motion. The ability to transform rotary motion from a motor into linear motion is used by many machines.

Another set of mechanisms focused on parallel motion—moving a link in space without rotating it. Parallel motion is useful for providing a precise grip for robotic hands, among other uses.

Finally, the above mechanisms shape motion. The shape of a motion is the path it follows through space over time.

If you are interested in mechanical motion, I highly recommend that you check out the cardboard kits and creations of Flying Pig, at www.flying-pig.com. As you go through your day, look at the machines around you and see what mechanisms they use to shape their motions.

Quiz

1. You have a stick. You drill a hole in it. What can you do with this?
2. You have TWO sticks and a wheel. What can you do with those?
3. What does a four-bar linkage do?
4. You want a wheel to go in one direction but not another. How do you limit it to one-way rotation?

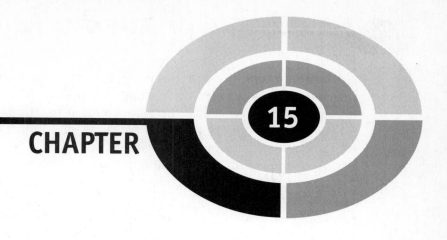

CHAPTER 15

Communication

Introduction

Robots aren't isolated machines thumping away quietly in a corner all by themselves. They may be part of a system of machines, or they may run in conjunction with human supervisors and operators.

Just as feedback is used to keep the parts of a machine under control, communication between machines keeps them working together to achieve a common goal.

Telerobotics

Not all robots operate alone, making parts in a factory or vacuuming your carpet. Some robots extend the reach of our human senses, giving our minds a way to operate in dangerous or distant environments where our bodies may not go. This is the field of *telerobotics*.

Telerobotics combines the ability to control the robot from a distance, *teleoperation*, and the ability to see, hear, and otherwise sense the environment around the robot, *telepresence*.

A remote robot may operate across vast distances, such as the explorer robot on Mars. It may operate across a short phyiscal distance but into very harsh environments, like cleanup robots in nuclear power plants or explorer robots in the mouth of a volcano. Or the robot may operate across large differences in scale, such as in microsurgical systems or small search and rescue robots.

Another term for telerobotics, teleoperation specifically, comes from science fiction. Robert Heinlein invented the term *waldo* for a remote manipulator, a mechanical extension of your body. Waldos may be electronically controlled, or they could be driven by direct mechanical linkages.

TETHERED ROBOTS

The simplest form of telerobot operates through a *tether*, which is simply a bundle of wires between the robot and its operator. You've seen these, and probably owned one yourself, in the form of a battery-powered toy car. The simplest form of control box is one or more switches that turn on motors in the remote "robot." Such a simple mechanism can barely be called a robot and, in fact, most of them do not have pretensions of being a robot. Mostly they are remote-controlled cars.

A tether is still a useful form of connection between the operator and the robot. It is more reliable and less expensive than other forms of remote control. In a more robotic case than the toy car, the operator gives high-level commands to the robot through the tether. The robot interprets these commands and carries them out to the best of its abilities, using its internal programs to work out the details of the operation.

Information comes back up the tether to the operator as well. The robot can send status and other information to the operator, as well as video, audio, or other forms of sensory information.

The operator becomes part of the robot control system, a high-level supervisor keeping everything on track. This adds another layer to our robotic control. At the lowest level, there is the basic control over individual motors and sensors. Above this there is one or more layers of computer programs that organize the individual motors, coordinating them into coherent patterns of motion. The program layer is a middle manager, interpreting the high-level commands and keeping the low-level systems organized. Finally, in a telerobotic system, the human operator gives commands to these programs, telling them what to do next.

REMOTE CONTROL

The next level of telepresence removes the tether cable and replaces it with light or radio waves. You actually experience telepresence nearly every day of your life. The television lets you see an environment that is far away in both space and time and, most often, reality. This fantasy world is broadcast to you on invisible waves and painted on your television screen with electrons. The telephone, or in my case the wireless telephone, gives you audio telepresence. Your voice is carried over wires and wireless mechanisms to be broadcast in a remote location, just as your ear can hear the sounds captured from this same remote location.

A telepresence link is not much more than a teleconferencing hookup, with pictures and sound. Along with these sensory signals there can be data signals, controls to the remote robot and status information coming back from it.

Today we have high-speed data communications in the form of DSL and cable modems, among other systems. We forget that not too long ago information was transmitted by simple sounds. If you wanted your computer to communicate with another, it would send a series of tones down the phone line using a modem (MOdulator/DEModulator). The other computer listens to these tones and turns them back into data.

You could listen in to this electronic conversation, and old-timers are very familiar with the "modem squeal." A noisy telephone line could turn this data into junk, of course, but that is always the case. Unclear communication, even on the human level, leads to loss of information. Many people still use tone-generating modems to access the Internet, and fax machines use a similar technology.

Remote control of a robot, then, could be nothing more than a modem connection over a wireless telephone to the robot. Of course, it could also be much more, depending on your needs and budget.

Semi-Autonomous

A remote-control robot is arguably not a robot at all, though there will definitely be argument. If a human is directly controlling the actions of the machine, is it a robot? If not, what is it?

If the human controller is simply giving high-level commands to the machine, more like a supervisor giving instructions to a worker, the machine starts to seem more robotic. At a certain point of abstraction the commands are less in the form of "turn left" and more in the form of "go explore that rock."

A robot that can operate by itself, under the instructive guidance of a remote operator, is said to be *semi-autonomous*. This is the twilight state between a remote-controlled machine and the fully *autonomous* (self-directing or, literally, following its own laws) robot.

There are times when a remote-controlled robotic telepresence is desired but not possible. When a robot is on Mars, for example, it takes between 4 to 24 minutes to communicate with it using a direct communication link. Light may be fast, but it's not infinitely fast, and Mars is a long way away. And communication to a robot on Mars is rarely direct or immediate. Communication normally occurs through several relays. The communication won't work if the planets are in the way, so there could be long delays while the planets slowly rotate back into position.

When it is impossible to directly control a robot, we turn to semi-autonomous systems. There are two control cycles in semi-autonomous systems. The high-level control occurs over a long time scale. A remote operator analyzes the most recent information from the robot and decides on its next course of action. This action is stated as commands to the robot—go here, do this, report on that. These commands are then sent to the robot.

Once the robot receives its commands, the second control cycle takes over. This is where the robot interprets the commands and tries to carry them out. It may have to deal with unforeseen problems, such as obstacles in its path or other unusual conditions. A clever robot can work around these obstacles, changing its path or avoiding problems as needed. The high-level commands are still carried out, but the details of their execution are up to the robot.

A robot that is capable of solving the small problems it encounters is able to follow its commands better than a dumb robot that only replays low-level commands to its motor control systems. The universe is messy and there are always little problems that get in the way of a robot's operation. When the robot can solve these problems without human intervention, that robot becomes more reliable.

As the robot carries out its instructions, it records information about its environment and internal status. This information is transmitted back to the control center, either during the operation or after the tasks have been completed. Note that the communication rate from Mars to Earth is pretty low, even slower than your desktop modem. It may take a long time to send information back about a short and simple task.

Once the operator receives this feedback, they can plan the robot's next actions.

Communication Technologies

After you have decided you want to communicate with your robot, the next question is *how*. Fortunately, there are many technologies available for robotic communication.

Data communication provides a way to send and receive numbers. These numbers, in turn, are collections of bits, or "on" and "off" conditions (Table 7-1). The numbers can represent anything the programmer desires. In terms of communication, we only need to worry about sending the bits. The meaning is assigned to them at a different level of the system.

PARALLEL

One communications system is the *parallel port*. Though something of a legacy device, having been mostly replaced by USB, most computers still have a parallel port on them.

When several activies are performed at the same time, they are said to occur in parallel. Thus, parallel programming involves two or more programs executing at the same time, and parallel communication involves multiple bits of data being sent at the same time.

The basic, plain-vanilla IBM PC parallel port is laid out as shown in Fig. 15-1. Note that there are two connectors displayed in this figure. The one on the left is normally found on the computer and is a "female" connector. The matching plug is, for hopefully obvious reasons, a "male" connector. I think that engineers probably spend far too much time in the company of other engineers and they really should get out more.

You communicate to the parallel port through three 8-bit registers. These registers are locations in the computer's memory that connect to the port. Writing a value to the register sends information to the port, and reading the register returns data from the port.

The data register sends eight bits of data at a time from the computer to the peripheral. The control register has four more bits that are used to change the behavior of the device. Finally, the status register lets the device send status information back to the computer. Printers are the most

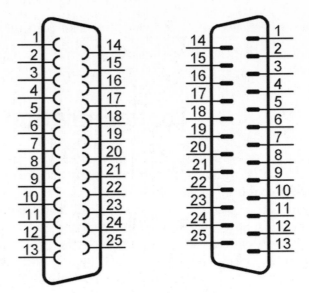

Fig. 15-1.   Parallel port

common peripheral for the parallel port, and the port definitions reflect this (Table 15-1).

Note that a bar over a signal indicates that the bit is inverted. If you write a 1 to the bit, it comes out as an electrical zero. Likewise, a 0 comes out as a positive voltage. The reverse is also true, a zero input to the port reads as a binary 1 to the program. The precise meaning of these signals and how to use them can be found in the reference named below.

A significant failing of the standard parallel port is that it can not transfer data efficiently from the device back to the controlling computer. Other parallel standards address this, however.

For more details on parallel ports, look to Jan Axelson's book *Parallel Port Complete* (Pub Resource, 1997).

SERIAL

While the parallel port sends eight bits simultaneously, the *serial port* transmits bits one at a time. Where the parallel port is an eight-lane highway, the serial port is a one-lane backwoods road.

In spite of being comparatively wimpy, the serial port is used more often than the parallel port. Part of its attraction may be that it uses fewer wires than the parallel port. Though the serial port defines nine pins in its interface

Table 15-1   Parallel port pins

Signal name	DB25 pin
*Data*0	2
*Data*1	3
*Data*2	4
*Data*3	5
*Data*4	6
*Data*5	7
*Data*6	8
*Data*7	9
Error	15
Select	13
Paper End	12
Acknowledge	10
$\overline{Busy}$	11
$\overline{Strobe}$	1
$\overline{AutoLF}$	14
Init	16
$\overline{SelectIn}$	17

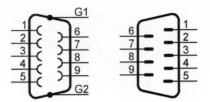

Fig. 15-2. Serial port

Table 15-2 Serial port pins

Signal name	DB9 pin
CD	1
RX	2
TX	3
DTR	4
GND	5
DSR	6
RTS	7
CTS	8
RI	9

(Fig. 15-2), you can communicate using just three of them, *GND*, *RX*, and *TX* (Table 15-2). The serial port is also bidirectional, in that it sends and receives data equally well.

Since serial ports are ubiquitous, almost all microcontrollers can communicate using the serial protocol. You can also buy sensors that send their information using serial communication.

This discussion of serial communication centers on the port found on most PCs, the RS232 serial port using a DB9 connector. There are actually two layers involved in using a serial port. There is the physical connection which involves the connector and the voltages applied to it, and there is the way these voltages are organized to transmit information.

The RS232 electrical structure can communicate over wires of at most 50 feet long. A different electrical system, RS485, can be used to send these serial signals over wires nearly a mile long. RS485 is an especially useful data carrier inside robots, since robots are electrically very noisy and the RS485 system is especially resistant to this noise.

There are also other specialized serial communications protocols. Your car probably uses the CAN serial interface, and in some systems the integrated circuits communicate using I^2C. There are other serial protocols, many others in fact, but they all send and receive data one bit at a time.

More information can be found in Jan Axelson's *Serial Port Complete* (Independent Publishers Group, 1998), as well as Edwin Wise's *Applied Robotics* (Delmar Learning, 1999) and *Applied Robotics II* (Prompt, 2002).

USB

The *Universal Serial Bus*, or *USB*, is a newer communications protocol that is beginning to edge out both the parallel and serial ports. It is more complicated than the serial port, but it supports all manner of peripherals, from computer mice to printers. As time marches by, more microcontrollers are supporting USB so we should be seeing it in more robots in the future. Needless to say, Jan Axelson has a book on the subject, *USB Complete* (Lakeview Research, 2nd ed., 2001).

TCP/IP

Your computer's connection to the Internet, or to your own home network, is probably made using the *Ethernet* networking technology and the *TCP/IP* (Transport Control Protocol/Internet Protocol) communications protocols. You can buy inexpensive microcontrollers that implement these systems.

Using Internet communications provides a way for your robot to communicate over existing computer networks. And yes, you should probably check out Jan Axelson's *Embedded Ethernet and Internet* (Lakeview Research, 2003).

WIRELESS

Once you have designed your robot using one of these communications techniques, you can buy a radio communications box to remove the wires between the robot and the operator.

There have been radio modems on the market for many years, and before them the ham radio operators had packet radio. These days wireless networks are big news, with 802.11b wireless Ethernet hubs and transceivers and their relatives. Using these inexpensive networks can give you an easy way to operate your robot remotely.

There are problems with most wireless systems, however. If you have ever carried a laptop around a wireless access point, such as a coffee shop or your living room, you may notice that the signal strength varies from one place to another. There will be dead spots where the network doesn't work at all.

If you are relying on your wireless connection to drive your robot, what do you do when the connection is broken? This is where semi-autonomous robots are useful again, because they are able to function without explicit control. Once the robot relocates a wireless signal, it can take on new instructions. When it is in a dead spot it can operate on its own.

Radio waves are not the only wireless communications technology. You can also talk to your robot using light, though this will only work if the robot can see the light.

Television remote controls use infrared (IR) light to send commands. The Mindstorms base station uses IR to program the robot controller. The RCX *Send IR Message* can send messages to another controller, while the *IR Message* sensor receives messages.

OTHER INTERFACES

Raw data is good for talking computer to computer, but there are times when you want to communicate with a robot directly. The traditional keyboard, mouse, and monitor provide a standardized way to interact with machines.

A more physical interface could be a control panel on the robot itself, consisting of lights, buttons, knobs, and dials. These components were common on old-time robots, especially old-time movie depictions of robots. A facsimile of a control panel can be generated using a modern LCD screen with touch-input capability, like those on palm-top and notepad computers.

Since you can buy a web camera with a microphone for about $50, you can give your robot eyes and ears. It is theoretically possible to communicate with a robot the same way as you communicate with another human, using speech and gestures. MIT has projects exploring these forms of communication with their Cog, Kismet, and other robots. Other universities are also exploring this territory to good effect.

The ultimate goal is for machines to interact with people as efficiently as people interact with other people—a goal that is still far in the future, but in just a few decades it has left the realm of science fiction and is now a very real possibility.

Academia is not the only place where advanced interfaces are created. Some games use cameras to watch the player, who controls their avatar in the game using gestures and other "body language."

Summary

There are many levels of communication in a robot system. Internally, there are feedback loops that serve to keep the mechanisms in balance and on track. At a higher level the control software manuals the operation of these low-level reflexes. Above that, the robot needs guidance and this is where the human steps into the picture.

Sometimes the robot is an extension of human senses and limbs, offering telepresence and teleoperation. Other times, though, there are technical reasons why simple telerobotics won't work so you have to build more flexibility and "smarts" into your robot, making it semi-autonomous. A fully independent robot could have value in some situations, but we normally want at least supervisory control over our creations.

Once it is decided that a robot should be remotely controlled, there are a number of technologies that can enable this control—from simple wires and switches to old-fashioned parallel and serial control protocols, to the newer USB and Ethernet.

The robot builder doesn't normally have to worry about the mechanisms of communication, since microcontrollers and computers tend to have one or more communication modules built in. You can even buy inexpensive chips that run TCP/IP over Ethernet.

The future of robot communication may be direct visual and vocal interaction with humans, though for now our communication is limited to data sent over wires, radio waves, and light beams.

Quiz

1. What role does the human operator have in the control loop?
2. What is a semi-autonomous robot? Why would you need one?

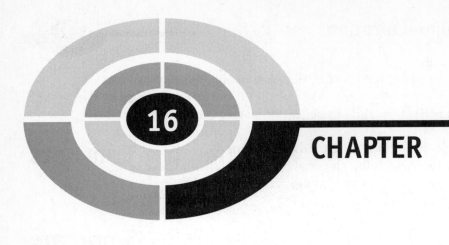

CHAPTER

16

Languages

Introduction

Language is how we communicate. It includes the complexities of our words and the ways they can be combined to carry meaning. Language is one of the most complex things we learn when we are little.

Communication is not just words. In human interactions, facial expressions, body posture, and the tone and inflection of your voice all carry meaning. We may talk about "body language," but is it language? We won't answer these philosophical questions here, but instead focus on language as it relates to computers.

Language—Computers
Any of numerous systems of precisely defined symbols and rules for using them that have been devised for writing programs or representing instructions and data.
Oxford English Dictionary, Online Edition

In Chapter 13 we looked at computer programming in terms of flow-charts and the RIS flowchart-based language. Those examples used graphical symbols as their language. In this chapter we map these symbols to the more compact word-based computer programming languages.

Programming Concepts

The programs in Chapter 13 used graphical symbols to represent instructions to the controller, "turn on the motor," "wait for three seconds," "read the light sensor," and so forth. These symbols were stacked together and the controller would execute one instruction and then move down to the next instruction.

These programs are written in an *imperative language*, where the *statements* (the computer equivalent of sentences) are commands to the computer. The connections between these statements define the order in which the statements are performed. They define the flow of control, making this a *control flow language*.

The fact that these programs are written with pictures and look like flowcharts makes this a *visual language*, but this is just a detail of presentation. Each program is ultimately reduced to a sequence of numbers. A sequence of words could just as easily represent these numbers. For example, the program in Fig. 13-7 (copied here as Fig. 16-1) could be rewritten as text.

```
BumperCar_1:
  do Initialize
  repeat (forever)
  {
    do Forward
    if Pressed(1) then
    {
      do TimedStop
      do BackLeft
    }
    else
    {
      if Pressed(3) then
      {
        do Timed Stop
        do BackRight
      }
    }
  }
  Initialize:
    PowerHigh = 4.0
    PowerLow = 2.0
    TimeBackup = 1.0
    TimeStop = 0.3
```

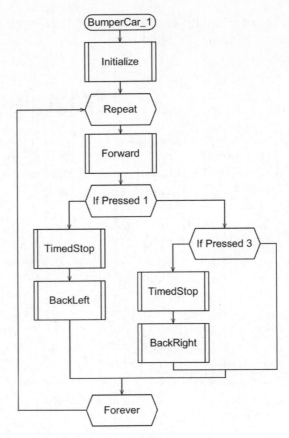

Fig. 16-1.   Sample program.

```
Forward:
  Power(A, C, PowerHigh)
  Direction(A, C, 1)
  On(A, C)

TimedStop:
  Off(A, C)
  Beep(2)
  Wait(TimeStop)

BackLeft:
  Power(A, PowerLow)
  Direction(A, C, -1)
  OnFor(A, C, TimeBackup)

BackRight:
  Power(C, PowerLow)
  Direction(A, C, -1)
  OnFor(A, C, TimeBackup)
```

The above is not any particular computer language but "pseudocode," a generic simplification of computer language. This is just one of many possible ways to represent this program, and not even a particularly realistic way. In spite of my having just made this pseudocode up, it clearly represents the same meaning as Fig. 16-1.

Finally, this program is *block structured*. The commands are organized in blocks of related statements. The statements between any two matching curly braces "{...}" are a block, as are the named sets of statements such as "BackLeft:." Block-structured programs map directly onto the shape defined by their flowchart.

Imperative languages follow the internal structure of the computer hardware, and they all have roughly the same types of instructions. There are instructions to change which instruction is going to be read next, instructions to put information into storage and to get it back out again, and instructions to test information from storage and perform different actions based on the result.

All programs can ultimately be reduced to a *universal Turing machine*. The Turing machine is a sort of primordial computer, created as an intellectual excercise by Alan Turing, a British mathematician and cryptographer. He was involved in the World War II Enigma code-breaking project, as well as the birth of modern computing.

TURING MACHINE

A Turing machine is both the most primitive computer it is possible to build and the second most powerful. It consists of four simple components: an infinitely long tape, a read/write head, a state register, and a transition table (Fig. 16-2, Table 16-1).

The *tape* contains an infinite series of cells that each hold one symbol from an alphabet. A simple alphabet consists of a blank (0) and a one (1). Each cell on the tape can be written to as often as desired, and of course you can read the symbol that is in a cell. All cells on the tape are assumed to hold a blank unless they have been explicitly set to a different symbol.

The *head* interacts with the tape. It reads the contents of the cell it is positioned over and can write a new value to that cell. It can also move the tape forward and backward, one cell at a time.

The *state register* is like a counter on the head. It keeps track of the current state, which may be a number or a name. A *state* is like a marker, a piece of memory that keeps track of where you are in the computation.

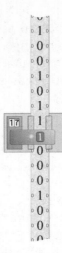

Fig. 16-2. Turing machine.

Table 16-1 Transition table

State	Read	State	Write	Move
1	1	2	0	R
2	1	2	1	R
2	0	3	0	R
3	0	4	1	L
3	1	3	1	R
4	1	4	1	L
4	0	5	0	L
5	1	5	1	L
5	0	1	1	R

The *transition table* contains the rules that drive the Turing machine. It defines a pattern to match, consisting of a state and a cell symbol. It maps this pattern to actions for the machine to perform. These actions consist of a new state for the machine, a symbol to write in the current cell, and a direction to move the tape. If there is no pattern matching the current state/cell, the machine stops and the program is complete.

Table 16-1 is an example transition table from Wikipedia. State 1 starts the process and the program execution is given in Table 16-2. The position of the head is indicated by the marked cell.

Table 16-2 Turing machine execution

Step No.	State	Tape cells								
1	1	0	1	1	0	0	0	0	0	0
2	2	0	0	1	0	0	0	0	0	0
3	2	0	0	1	0	0	0	0	0	0
4	3	0	0	1	0	0	0	0	0	0
5	4	0	0	1	0	1	0	0	0	0
6	5	0	0	1	0	1	0	0	0	0
7	5	0	0	1	0	1	0	0	0	0
8	1	0	1	1	0	1	0	0	0	0
9	2	0	1	0	0	1	0	0	0	0
10	3	0	1	0	0	1	0	0	0	0
11	3	0	1	0	0	1	0	0	0	0
12	4	0	1	0	0	1	1	0	0	0
13	4	0	1	0	0	1	1	0	0	0
14	5	0	1	0	0	1	1	0	0	0
15	1	0	1	1	0	1	1	0	0	0
16		*halt*								

The most powerful computer is a *universal Turing machine*. In this universal machine, the content of the tape is expanded so that it describes the transition table of a Turing machine as well as the program it runs. The universal Turing machine then simulates this described machine.

Any language that has all of the capability of a universal Turing machine can emulate any other computational machine. Such a language is known as a *Turing complete* language.

A real computer language can, in fact, be as simple as the Turing machine. The rudely named language that we shell refer to as BF is one such. It's more an excercise in what is possible than a practical language. It has only eight instructions:

> \>   Increment position
> \<   Decrement position
> \+   Add 1 to the value at the current position
> —   Subtract 1 from the current value
> .   Output the current value as a character
> ,   Input a character from the keyboard and store it as the current value
> [If the current value is zero, jump to the statement after the next "]"
>]   If the current value is not zero, jump to the statement after the previous "["

BF programs are, essentially, unreadable strings of characters. For example, the Wikipedia sample program that prints "Hello World!" is:

```
++++++++++[>+++++++>++++++++++>+++>+<<<<-]
>++.>+.+++++++..+++.>++.<<+++++++++++++++.
>.+++.------.--------.>+.>.
```

For more details, Google for "BF Language."

Choosing a Language

Since all computer languages can be reduced to a Turing machine, it doesn't matter much to the computer what language you actually use. However, nobody in their right mind is going to try to write a word processor in BF. Just because something can be done does not mean it should be done.

There are many different programming languages to choose from, though not all of them are usable for all applications. When possible, you should pick a language that works well for the problem you are trying to solve.

Different computer languages package their capabilities in different ways, trying to make a particular type of job easier. The Robotic Invention System language, RCXScript, has handy statements for robot control like "Set Direction" and "On" and "Beep." These statements make it easy to drive your robot around. Other languages try to make it easy to create large applications (C++, C#), or create programs that can run on many different computers (Java), or to make small, fast programs (Assembly, Forth).

The language you program in is determined by the languages that you know, so it is good to be familiar with many different computer languages. Then you can pick the best one for a project instead of trying to shoehorn the one you know into doing all of your projects. Even on a computer as simple as the Mindstorms RCX controller, there are several language choices available.

Many popular computer languages are based on the C programming language. So it is no surprise that there is a version of C for the RCX, running on Markus Noga's BrickOS which is, essentially, a new personality for the RCX. The object-oriented C, C++, is also available on BrickOS. Dave Baum has a version of C called Not Quite C (NQC) that runs on the RCX brick as shipped by LEGO. Java runs on LeJos, and there is a form of Visual Basic that runs on SpiritOCX. Not to mention pbForth. There is no shortage of languages to learn. This chapter is just a teaser, touching on a few highlights, since there is not enough room to go into much depth.

While some languages are definitely only suitable for certain types of tasks, the languages listed above are general in their application. Whether you use C++, Java, or the Visual Basic interface may come down to which one suits your own personal style.

If you find you are happy with the RCXCode visual language that comes with the kit, but would like to be able to do more with it, National Instruments has a souped-up visual language called RoboLab. This is sold through the educational branch of LEGO, Pitsco-Dacta. It is a *dataflow language*.

In dataflow, you have statements or icons that represent actions. But instead of linking them together by order of execution, you link them together by the data that they pass to each other. Each statement executes when all of its inputs have the data they need, and when the statement is done it may send more data down a link to another statement.

A third type of language, the *logic languages,* are *declarative languages*. They don't have statements that translate into actions at all. Instead, they define relationships in the form of mathematical or logical equations. Then, given a set of data, the language determines the best way to fulfill the

relationships. Because many real-world applications require some verbs, the declarative languages can include "hooks" where actions are specified.

MANUFACTURING LANGUAGES

One type of robot is the computer-numerically controlled (CNC) machine. These are computer-driven power tools that can cut, burn, punch, fold, and, if incorrectly programmed, mutilate your sheets and blocks of wood, steel, marble, cloth, and aluminum.

CNC machines have their own control languages. Though there are many different forms of machine control language, one of the most common is the GM code. GM codes, and other CNC code systems, are not complete computer languages. They are not Turing complete and only support the most basic operations.

The G codes are geometry codes that direct the machine to move its cutting head. For example:

G00 X Y Z   Move rapidly to position X,Y,Z
G01 X Y Z   Move at cutting speed to position X,Y,Z
G02 X Y I J   Cut a clockwise arc to position X,Y around center I,J
G03 X Y I J   Cut a counterclockwise arc

The M codes are machine codes that control the operating state of the machine. For example:

M00   Stop the machine
M17   Cutting torch on
M18   Cutting torch off
M75   Enable punching tool
M85   Disable punching tool

Each machine will have its own unique set of geometry and machine codes depending on its abilities. The more advanced robot arms have more complicated codes.

CUSTOM LANGUAGES

Sometimes it is convenient to create your own custom "mini" language. These lie somewhere between GM codes and full-scale programming languages in complexity, and may or may not be Turing complete. Most computer games have their actions programmed by specialized scripting languages, for example.

Though you have to go through extra work to create a custom language, it will closely match the needs of your problem. You then save time solving the problem itself. If you have to solve this type of problem several times, you end up saving time in the long run.

HUMAN LANGUAGE

There has been a continuing effort in computer science to try and program computers in human or "natural" language, or at least something close to natural language. This is tricky since human language is filled with subtleties, multiple meanings, intricate contexts, and hidden assumptions. The computer, on the other hand, requires detailed and explicit instructions that guide it down to the smallest detail. The gulf between these two forms of language is huge.

While trying to program computers in natural language has not been very successful, there has been good work done in getting computers to at least make sense of natural language. The programs we have used so far are commands of action: turn on this motor, read that sensor, decide on a course of action based on a variable.

Programs can also be used as translators, mapping from one system of symbols to another. When one of those systems of symbols is human language, the program has to work very hard indeed. If it is spoken language, it is even harder.

This difficulty in understanding language is why almost all robot and computer interfaces consist of keyboards used to enter simplified commands, touch-screens to select from multiple options, dials, knobs, and lights. These are the mechanisms that the machine understands.

HUMAN INTELLIGENCE

If it is hard to map human language into computer instructions, it is even harder to map human intelligence into the computer. What does it mean to think, and how do we describe that in terms the computer understands?

Artificial intelligence is one of the loftier goals of computer science. Though the intelligent, nearly-human robot is a popular figure in science fiction, the reality is still a long way away.

Some of the more interesting attempts at mapping biological intelligence onto the abstract symbols of mechanical computing take inspiration from the neuron itself. We explore the topic of neural computing in the next chapter.

Summary

We took a quick look inside the world of computer languages. At their most abstract all languages do the same thing, as modeled by the universal Turing machine.

Given that computers don't really care how you program them, you can choose a language that works well for the problem you are trying to solve.

Fortunately for the programmer, almost all computing environments, even the Mindstorms RCX, have a multitude of languages available to them. There is no one "best" language, but each language has its own strengths and weaknesses.

Quiz

1. What is communication? Name a few forms of it.
2. How does the computer see a program?
3. Describe the Turing machine.
4. What are three types of computer language?

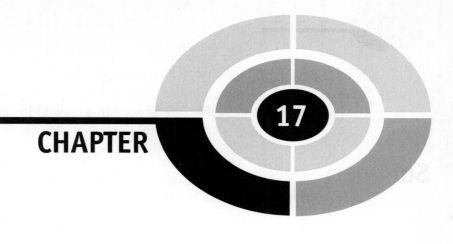

Intelligent Behavior

Introduction

When you think about robots, the first thing you may think about is the behavior of the robot, how it acts and reacts. The different types of robots diverge here, in the discussion of behavior. Most robots have the same needs when it comes to their mechanics and electronics. But different kinds of robots have different needs when it comes to behavior, especially intelligent behavior.

Even the use of the word *intelligence* leads to difficulties. What is intelligent behavior for a milling machine? Do you really want your punch press to get bored or restless as it's cutting out its millionth part? For most everyday robots, you want reliable, repeatable, and consistent *behavior*—not human, or even animal, intelligence.

"Intelligence" only comes into play for advanced and, usually, mobile robots. You want your Mars rover to have a measure of intelligence. A robot that acts as a tour guide, or any robot that interacts with people, needs to have some intelligence. Researchers are developing ways to add intelligence to these kinds of robots.

Even then, intelligence is not necessarily easy to define. Is a chess-playing robot intelligent, or just really good at solving board positions? How about a visual inspection robot? Is intelligence the ability to think logically? Programs can do logic, they are quite good at it. But they are bad at things that we don't normally consider a part of intelligence, such as recognizing faces.

Our friend the OED defines intelligence as "the faculty of understanding". Think for a minute, what does it mean to understand something? What do you know about a pencil, for example? How did you come to know it? This chapter actually deals with behavior more than intelligence.

Reflexive Control

Most robots have a layer of purely reflexive control. These *reflexes*, which are simple actions that occur without thinking, are all that is needed for some robots. They provide a form of *homeostasis*, literally, resistance to change. Homeostasis is a complex response to stimulus used to keep a system within acceptable limits. Your body's response to heat, sweating, is such a response. Of course, your home thermostat is also practicing a form of homeostasis.

Figure 17-1 shows a way at looking at this low-level control. The system we are considering is contained in the dashed box. We receive a command outside of this system that defines a goal state, such as a particular temperature or a position in space. Also outside of the system are the unknown and unknowable disturbances that affect the robot. Inside we have the robot,

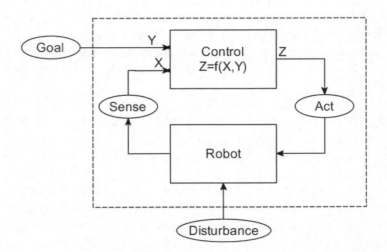

Fig. 17-1. Reflexive control.

or other mechanical or chemical system, and its control system. The control system senses the state of the robot and uses this information, in comparison with the goal, to generate output actions.

The equation $Z = f(X, Y)$ means that the output action Z is a function of the sensed state X and the input goal Y. The function can be as simple as equation (17-1):

$$f(X, Y) = Y - X$$
$$Z = Y - X$$

$$(17\text{-}1)$$

If the sensed state is "greater than" the goal, the action will be to "increase" the state of the robot so it matches the goal. Hardly rocket science.

This only works with numbers, of course. The goal, the sensed state, and the action are all assumed to be represented by numbers. Equation (17-1) makes the assumption that the output action directly affects the input sense, using the same range and scale of values. The real world is rarely so accommodating.

THERMOSTAT

The traditional control example is the thermostat. The goal is a temperature you set and the sense is the actual temperature. The output is not, however, a power setting for the heater but instead a simple on/off switch. Most home heaters operate at full on or full off.

At first thought, the control logic might look like:

```
If temperature is greater then setting then turn heater off.
Otherwise, turn heater on.
```

This is a terrible control system. The heater will be clicking on and off all the time.

To fix it, we introduce our old friend hysteresis. By defining a *dead band* around the desired setting we can reduce the number of on/off transitions (Fig. 17-2). The control logic is only complicated a little bit by this hysteresis. Note we are using a compact notation here, just one of many possible notations:

$$b = \frac{DeadBand}{2}$$
$$Z = 1: X < (Y - b)$$
$$Z = 0: X > (Y + b)$$

$$(17\text{-}2)$$

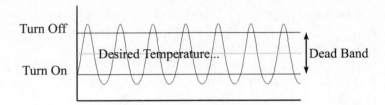

Fig. 17-2.   Dead band.

where b is one half of the dead band. Z is set to 1 (the heater is turned on) when X (the temperature) is less than Y (the desired temperature) minus b (half of the dead band). Likewise, if the temperature goes above the dead band's top, turn off the heater.

> *There is a reason we jump around, illustrating concepts using different notations. The symbols used to communicate an idea are not as important as the idea itself. There are many different ways to say any given thing. The more of these ways you can recognize, the better. While it may be confusing at times, it builds character.*

Note that when the heater is turned off, the environment continues to get hotter for a while. Also, the temperature falls a bit after the heater has turned on. Physical systems have momentum, a tendency to continue in their current "direction." The amount that the measured value goes above the target value, in this case the top of the dead band, is called *overshoot*. Likewise, falling under the mark is *undershoot*.

PID MOTOR CONTROL

A more interesting example, and one dear to the heart of robot builders everywhere, is the ability to control the position of a motor-driven mechanism. In this example, both X and Y are in terms of a physical position and the control output Z is the power applied to the motor. Since Z is not just an on/off switch we have more subtlety available in this controller.

Ignoring a few complexities for a moment, we can expand on equation (17-1):

$$Z = K_P \times (Y - X) \tag{17-3}$$

This, and equation (17-1), are *proportional controls*, since their outputs are in proportion to the error $(Y - X)$. The new factor K_P is a gain control. It lets us

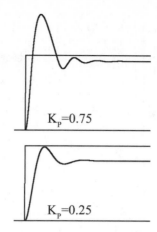

Fig. 17-3.   Proportional control.

adjust how quickly the mechanism moves toward the goal. Figure 17-3 shows this formula in action. The square pulse is the target value, while the wiggly trace is the actual value of the system under control.

The top trace has a proportional gain of 0.75 and shows a fair amount of overshoot. The oscillation that follows is called *ringing*. After this, it settles down into a steady line a little bit under the set point. The bottom trace has a smaller gain of 0.25 and avoids the overshoot, but it's settling point is a bit lower than the set point.

Larger proportional gains settle closer to the set point but have more overshoot and more ringing. Smaller gains don't overshoot, but then they never really reach the goal either. If there were only a way to get rid of that offset. . . .

Instead of using just the current error in position, we could accumulate, or add up, the error over time. This way, even a small offset would add up over time and could move our mechanism closer to its goal. This is shown in equation (17-4):

$$e = (Y - X)$$

$$e_I = e_I + e \qquad\qquad (17\text{-}4)$$

$$Z = K_P \times (e + K_I \times e_I)$$

This describes proportional-integral control.

We broke the $(Y - X)$ error term out into its own symbol e. Then we keep adding our instantaneous error e to the accumulated, or *integral*, error e_I. Adding the new integral gain K_I lets us control how important this integral error is to the control. K_I needs to be kept small to avoid massive ringing

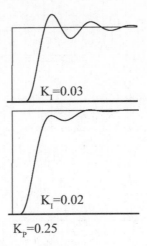

Fig. 17-4.   Proportional-integral control.

(Fig. 17-4). In both traces K_P is set to 0.25. Note the extra ringing in the top trace with K_I set to 0.03. The subtly smaller K_I of 0.02 reaches the goal with a minimum of fuss.

It is, however, slow in reaching the goal. Is there a way to make the response a bit snappier? Actually, having to react to a massive change in goal is unusual. Goal states are more likely to change slowly. In Fig. 17-5 we see the PI controller following a smoothly changing signal.

There is one more term, a *derivative*, that is used when there is a time delay between the action output and the sensed input. This term tries to predict the future based on the error history:

$$e_{t-1} = e_t$$

$$e_t = (Y_t - X_t)$$

$$e_I = e_I + e \qquad (17-5)$$

$$e_D = e_t - e_{t-1}$$

$$Z = K_P \times (e + K_I \times e_I) + K_D \times (e_D)$$

We introduce time into the equation, where t is the current time and $t - 1$ is the previous time slice. This completes the control, creating a *proportional-integral-derivative control* or, more simply, a *PID control*.

This is just one, simple, form of the PID math—there are others that give better results at the expense of more calculation complexity. The PID needs

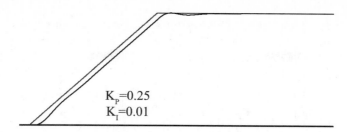

Fig. 17-5.   PI control with gradual change.

to be tuned to match a specific problem. While the formulas are reasonably generic, the gain factors need to be adjusted to get the best control.

Now for one of those complexities that we ignored earlier. What are the units we are generating for Z? Right now, Z is in terms of the input. If the inputs X and Y were scaled to the range $[0...1]$ by dividing them by their maximum values, the output is then in terms of percent power. This can then be converted into a power level for a motor or other mechanism.

Serial Behaviors

A PID or other reflexive control is fine for the most low-level tasks. There are times, however, when you want your machine to perform a series of actions, one after the other. Industrial robots don't need intelligence, per se, but do need to have predictable behavior. These robots will consist of some low-level reflexes and a bunch of simple commands to execute, one after another.

For another example, in a competition you need to might need to follow a path for a certain distance, retrieve an object, and then return to the start. A Mars explorer may be given a set of tasks by the ground control and it must then carry them out, one at a time, until it is done. A smart Mars explorer would even be able to skip a task if it found itself unable to complete it.

There is not much to say on the topic of serial behaviors, though there are some subtleties to be aware of when implementing them. The first is, how does the robot know when to stop one behavior and start the next behavior? You could do something simple like "drive forwards for 10 seconds" or "turn this wheel 100 times," but those are special cases that are only useful in a controlled environment.

If you are a Mars explorer, how do you know when you are next to that funny-looking rock? If you are driving over rough or slippery ground, how do you know when you have gone 100 feet? Some of these questions might be answered by specialized sub-programs that can recognize such a poorly defined situation. This is difficult, one of those hard to define tasks like recognizing faces. Animals do it all the time, robots have to work really hard on the problem.

Even when you have a situation where it is easy to tell when to switch tasks, such as when you have crossed a black line on the ground, you have bumped into a wall, or perhaps wandered into a shadow, you still don't know *exactly* where you are. You may have crossed a line, but at what angle? How far did you roll over it before you came to a stop? If the next step in the task relies on the robot being in a particular position or situated next to something just right, you will probably need a bit of code to align the robot before it really transitions.

Everything a robot, or any other machine, does has error in it. Some errors are well behaved, like the backlash of gears. The error stays within a known range and doesn't grow over time. Other kinds of error get worse over time. If you have a slipping drive belt between a motor and actuator on your robot, that slippage will make the error in position grow worse over time.

Mobile robots have this problem no matter how they are built. The interface between their wheels (or treads, or legs, or whatever) and the ground is not mathematically perfect. Any calculations used to determine the robot's position based on what the wheels have done are going to have errors. Every time the robot turns or moves, a little more error creeps into the picture.

More feedback can be used to reduce the error. Distance sensors, cameras, bump switches, GPS, anything that can interact with the outside world, can provide a second opinion on the robot's position. When you have more than one way of estimating the robot's position, you can combine the results to reduce the error of any one technique. This sensor fusion is not easy, but at least it is possible.

Layered Behaviors

The opposite of serial behavior would be parallel behavior. But you can't really have a bunch of behaviors active all at the same time, at least not if they are driving the same robot. What you need is a way to layer these behaviors, activating the right one at the right time.

If you've been keeping up, you will notice right away that the big challenge is knowing when the "right time" is to activate a behavior. It could be when

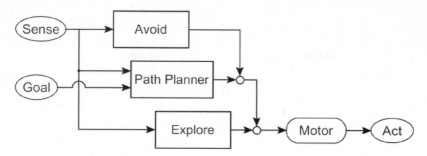

Fig. 17-6.   Subsumption layering.

the battery is low, or when the robot has bumped into a wall, or when a command is given to the robot. Maybe the robot needs to seek out some person and deliver a message. Some of these conditions are easy to detect, others are harder.

Assuming that each behavior is able to determine for itself when it should be active, these behaviors could be layered. Figure 17-6 shows three possible navigation behaviors for a mobile robotic explorer.

When there is nothing else happening, the robot is free to explore randomly. If it has been directed to go somewhere specific, the path planner takes control from the explore module. The robot then travels to the destination. If, once it gets there, the path planner releases control, the robot starts to explore again. If the robot finds itself in trouble, the emergency avoidance module takes over until the robot is in the clear again.

The different behaviors override, or subsume, each other, making this a *subsumption* architecture. This approach to merging multiple simple behaviors into a cooperating, complex whole was invented by Rodney Brooks, who is currently the director of the MIT Computer Science and Artificial Intelligence Laboratory.

The complete subsumption system allows behavior modules to override inputs as well as outputs, and to reset other modules. Extensions to the basic system provide global information, representing hormones and other systemic indicators, to help control the overall behavior.

Most modules in a subsumption system are *reactive* behaviors. These are like reflexes, though a bit more complex than the reflexive PID controller shown above. A reactive behavior doesn't do any planning, but reacts to its sensory input. Layers of reactive behavior provide an excellent middle layer in robotic control, giving the robot complex behaviors from small, easily programmed pieces. These, in turn, give the robot the ability to react smoothly and quickly to its environment.

Logical Behavior

While mindlessly roaming around an environment is good for theoretical robots and, perhaps, vacuum cleaners and mine-clearing robots, other applications require a more structured approach. Something that provides the *Goal* from Fig. 17-6.

It may seem like the best way to provide high-level control for your robot is to just write code to do what you want. This is a *heuristic* approach. A heuristic is a solution to a problem that solves the problem, but doesn't guarantee that the solution is the best one. It just gets the job done.

In the long run, an undisciplined approach causes more problems than it solves. If the system is going to have a long life it pays to put more effort into the control system.

SCRIPTING

As mentioned above, many robots don't need intelligence, they just need some coherent guidance. In this case, a scripting language works perfectly well to control their actions. GM control codes, mentioned in Chapter 16, are a form of scripting language.

For fancier robots, like robot arms, the control gets more complicated. A script would have to specify three-dimensional position in space, plus information about the angle of the end actuator. This is six axes of information, *X*, *Y*, *Z*, and pitch, roll, and yaw (angles around the axes). If there are obstacles to avoid, that adds to the complexity.

Once you have decided on a position and rotation in space, the lower-level controller has to calculate the exact angles all of the joints must take to reach that position. This is the process of *inverse kinematics* and requires some heavy-duty mathematics.

FORMAL LOGIC

Computer programs are full of code that makes use of Boolean logic. "If this condition is true AND that condition is true, perform this action." The basic Boolean operators are AND, OR, and NOT, though the exclusive-or XOR is also popular. The Boolean operators take one or more inputs and return a single output value. The inputs and outputs are all True/False values, Boolean binary. We use "0" for false and "1" for true.

The easiest way to describe the operators is with a *truth table*. Table 17-1 shows this for the basic Boolean operators. For example, the AND operator only returns true if all of its inputs are true. OR returns true if any of its inputs are true. Complicated conditions can be built up, specified by a truth table. Any arbitrary truth table can then be decomposed to an equation using the basic operators of AND, OR, and NOT.

More interesting results can be found using predicate logic. Using predicate logic, and its various relatives, you can represent facts in the computer's memory. From these facts, the computer can infer new facts or find new relationships between existing facts. For example, given these assumptions:

Rabbits are faster than turtles
Turtles are faster than snails
Robert is a rabbit
Steve is a snail

You could ask the system if Steve is faster than Robert and it would answer no. Nowhere is there a rule that says Robert is faster than Steve, or even that Rabbits are faster than Snails. The information provided is sufficient to figure out the answer.

Logical programming was the basis of artificial intelligence when the field first began, and it still has lots of value in expert systems and some other formal information-processing fields. It doesn't relate directly to robot control and it doesn't really represent how people think.

NATURAL COMPUTATION

Artificial intelligence researchers and robot builders alike have been looking to nature for inspiration on how to build their systems. All of the animals have successful control systems, providing quick reactions to a changing environment as well as high-level planning and guidance—the "thinking meat" of Terry Bisson's short skit, *They're Made of Meat*:

   . . .
   "Oh, there is a brain all right. It's just that the brain is made out of meat!"
   "So what does the thinking?"
   "You're not understanding, are you? The brain does the thinking. The meat."
   "Thinking meat! You're asking me to believe in thinking meat!"
   . . .

Fortunately for us, our thinking meat serves us well.

Table 17-1   Boolean operators

AND		
A	**B**	**Out**
0	0	0
0	1	0
1	0	0
1	1	1

OR		
A	**B**	**Out**
0	0	0
0	1	1
1	0	1
1	1	1

XOR		
A	**B**	**Out**
0	0	0
0	1	1
1	0	1
1	1	0

NOT		
A		**Out**
0		1
1		0

The lowly neuron is at the core of all of the brains that we have examined so far. The *neuron* is like a little computer itself, though it seems to perform fairly simple computations. A bunch of neurons together, a *neural network*, appears to work magic.

While we don't know how the human, or any other, brain works in detail, that doesn't keep us from using what we do know to design more effective software. One of the things that brains do best is to make sense of incomplete information. A form of pattern matching—finding the pattern behind noisy and incomplete data. Recognizing faces. Those tasks that traditional computer software finds really difficult, brains do with ease.

Pattern Recognition

One of the common problems noted in earlier sections was the problem of knowing when a task is done. This is a *pattern matching* or *pattern recognition* problem. Have you found the right person? Are you in the right part of the room? Has the robot's arm reached the target yet?

There are many approaches to pattern matching. For simple sensors a trivial test may be sufficient. "Is the reflected light below 30?" Normally, though, there will be more than one sensor. There is often more than one value associated with a single sensor. Cameras, for example, return thousands if not millions of values, the individual pixels in their image.

STATISTICS

Statistics is the branch of mathematics that is concerned with possibilities. Statistical methods can assign a value to how likely something is. For example, given a half-dozen distance sensors around your robot, you should be able to calculate the probability that you are in a corner, or between two walls.

Bayesian analysis is particularly relevant to the pattern-matching problem. In daily life, you are most likely to find Bayesian techniques in your e-mail SPAM filters. Statistics, however, is a complex and arcane field that we cannot do justice to here.

FUZZY LOGIC

Fuzzy logic is the insecure cousin of Boolean logic. Boolean thinking is crisp and precise, definite. The temperature *is* hot or the temperature *is* cold. Fuzzy logic allows the temperature to be *mostly* hot but still a *little bit* cold. Boolean logic can be seen as a graph like Fig. 17-7. The moment the temperature passes the 80 degree mark it is suddenly 100% hot. Pow!

Figure 17-8 shows the same thing from a fuzzy perspective. The degree of "hotness" gradually increases as the temperature goes from 70 to 90 degrees. At 80 degrees, it is about half hot and half cold. Warm, if you will. If you are from any of the Northern states, substitute 70 (or even 60) degrees for our 80 degree Texas definition of warm.

When you think in Boolean logic, everything switches 100%. "If the temperature is Hot, turn on the fan." On or off. This works well for some environments and it is simple to implement. In fuzzy logic you think in a sliding scale. "To the degree that the temperature is Hot, turn on the fan." If the temperature is only a little bit hot, say 75 degrees, the fan turns on low. If the temperature is 100 degrees the fan is spinning at full tilt. This varying output value, from not hot to fully hot, is based on the input's *membership* in the category "hot." There are a number of membership functions, such as those shown in Fig. 17-9.

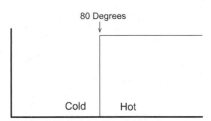

Fig. 17-7. Boolean hot or cold.

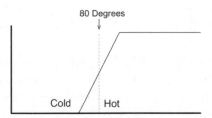

Fig. 17-8. Fuzzy hot or cold.

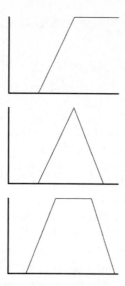

Fig. 17-9. Membership functions.

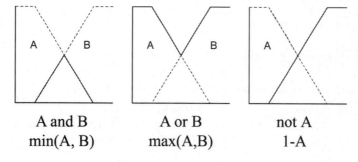

Fig. 17-10. Fuzzy operators.

There are fuzzy versions of the Boolean operators, too. Instead of a truth table, we represent them as equations and pictures (Fig. 17-10). The *min( )* function takes the lower of both values, while *max( )* takes the larger.

Fuzzy logic also has some modifiers that affect the shape of the membership function. If you want to know if something is *very* hot and not just *hot*, you could shape the "Hot" membership function using the "Very" modifier before you check the temperature against it.

Some modifiers, also known as *hedges*, are shown in Fig. 17-11. These assume that the output value for 100% membership in a category is 1.00. The hedges take the membership function's points and raise them to a power. Powers greater than 1.0 tighten the function, making it harder to be a full

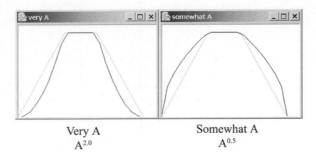

Very A
$A^{2.0}$

Somewhat A
$A^{0.5}$

Fig. 17-11. Fuzzy hedges.

member of the category. Lower powers bloat the membership function, making it easier to qualify.

Fuzzy logic is a simple form of pattern matching. It classifies its input into membership categories. These membership values can then be used to weight the actions associated with that category. If several categories are trying to drive the same output, their conflicting control values can be averaged together, using the priorities defined by their membership weights.

NEURAL NETWORKS

Computational neurons, and the collections of neurons known as neural networks, are pattern-matching modules based on biological neurons. Each neuron takes one or more inputs that it compares against its inner template. The output of the neuron is then active to the extent that the input matches that template.

Figure 17-12 shows a diagram of a typical computational neuron, known as a McCulloch-Pitts neuron. Though a bit scary-looking, its operation is simple. The input is a vector, or list, of numbers called x, each entry of which is indexed as $x(i)$. Internal to the neuron is a set of weights w, one for each input. The sigma operator 'Σ' is a loop that adds up the results from its equation, which is calculated once for every input index. If there are three inputs, the equivalent equation is:

$$y = (w(1) \times x(1)) + (w(2) \times x(2)) + (w(3) \times x(3))$$ (17-6)

y is the output value.

This function is called a *dot product*, and it is one method of calculating how closely two vectors match. It also has further mathematical significance, not relevant to this discussion.

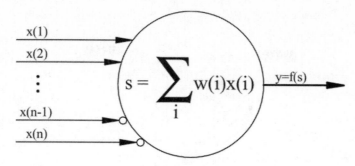

Fig. 17-12. McCulloch-Pitts computational neuron.

For the dot product to give a result between 0.0 and 1.0, both the input vector and the weight vector need to be *normalized*, so they each have a length of 1. What does it mean for a vector to have a length? The list of numbers in the vector represents one end of a line in an unusual multidimensional space. The other end is at the origin, where all values equal 0. The length is calculated as:

$$L = \sqrt{\sum x(i)^2} \qquad (17\text{-}7)$$

You will recognize this as a generalized version of the equation for a line's length, or a distance measurement, in three-dimensional space:

$$L = \sqrt{x^2 + y^2 + z^2} \qquad (17\text{-}8)$$

To normalize the vector, divide each element of the vector by the vector's length:

$$x(i) = \frac{x(i)}{L} \qquad (17\text{-}9)$$

where L is the length calculated in equation (17-7).

The ability to normalize a set of values is handy. Normalization scales things to a consistent range of values, and makes a number of mathematical operations behave better. All of this is a fancy way of testing a list of input numbers against a template, and then returning a value that indicates how closely they match.

Several neurons, each with its own template, can be watching the same input values (Fig. 17-13). The output from all of these neurons will probably show that one of them is the best match against the input. This result can be used to trigger an appropriate action.

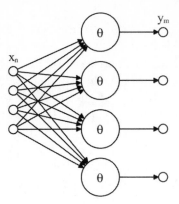

Fig. 17-13.   Simple neural network.

Note that the set of outputs creates a pattern that could itself be interpreted by one or more additional neurons. A second layer of neurons abstracts the result. It is like the eye seeing a pattern of colors and recognizing it as fluffy. The next layer might abstract that into the concept of *cat*, and the next might know it as a *mammal*, then an *animal*. (This example oversimplifies a complex problem, and does the neural network something of a disservice by anthropomorphizing it.)

Given these simple computing units, the field of neural networks proceeds to find various ways to connect them to give useful results. While the patterns in the neurons can be set by hand, it is often more useful for the neural network to be trained against example data, or even self-trained by turning it loose in its environment. Networks that have been trained against examples can find patterns in the data, perhaps even patterns that the programmer would not have noticed. If trained correctly, the networks are also resistant to noise. The information coming into the network can be incomplete or even partially wrong, and the network will be able to recognize it.

Summary

This chapter provided a brief tour through several levels of behavioral control. At the bottom of the intelligence scale were the reflexive controls, exemplified by the PID industrial control formula. These reflexes use sensory feedback to keep the lowest level of the robot, the motors and other actuators, on track.

Different behaviors and reflexes can be combined and coordinated at the middle level. The simplest form of coordination is the simple serial, scripted manager that selects behaviors one at a time until the task is complete.

The subsumption architecture provides a parallel approach, allowing the different behaviors to override each other as needed, or as triggered by their environment. This architecture is a useful middle-level control for mobile robots.

At a higher level of abstraction, formal logical systems can provide a framework to manage knowledge. While this appears to be a bit too abstract for robot control, it can have a place. Can you think of a way that logic can be used by a robot?

One of the harder problems to solve in robot control is recognizing a particular object or situation, so the robot can change its behavior appropriately. Heuristics, statistics, and natural computation in the form of neural networks are all technologies that address this pattern-matching problem.

Quiz

1. Define intelligence. Is intelligence good for a punch press machine?
2. What does a PID control used for? What does it do?
3. For sequential behaviors, what is the tricky part? How about for layered behaviors?
4. Describe a pattern-matching system inspired from nature.

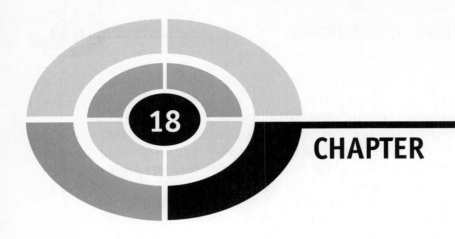

18

CHAPTER

Advanced Control

Introduction

The subject of robotic control is huge. You could fill a small library with the research and control techniques used to manage both industrial robots and the more ambitious research robots, many of which are trying to fill the role of a human. Some of these humanoid robots are caretakers, helping out the disabled. Others are guides—either mechanical or virtual (software only) tour guides for museums, hospitals, or online help systems. Then there is NASA's humanoid robot designed to face the rigors of space. The nonhuman explorer robots, wheeled rovers or legged explorers, need the flexibility of intelligent control as they wander harsh or distant landscapes.

This chapter is a continuation of Chapter 17. In the previous chapter we focused on control systems and pattern matching to enable the robot to switch between behaviors as needed. This chapter looks at decision-making a little bit more and then spends the rest of its time exploring new topics.

One of the key aspects of intelligence as we recognize it is the ability to learn. We look at some apparently simple learning tasks and then take a deeper look at techniques for robotic education.

You may have noticed a distinct shortage of hands-on projects in these control chapters. While it is easy to snap together a mechanism to illustrate a mechanical concept, or wire together a few components to show off a circuit, control programs are harder to create.

While the RCX flowchart language provides for some easy examples, it is also limited in what it can do. The other text-based languages mentioned in Chapter 16 have the power we need, even on a simple controller like the RCX, but involve a lot of infrastructure to make them work.

With programming, you enter into the world of complex grammars and the supporting network of tools needed to make the languages work: compilers, downloaders, debuggers. It takes several chapters just to get started with programming. So most of our explorations of control concepts are in the abstract. I encourage you to look up more books on the subject of Artificial Intelligence, the foundation of most nonindustrial robotic systems. For example, *Hands-On AI with Java* by Edwin Wise (McGraw-Hill, 2004) provides a better introduction to the subject.

Decisions

We have already peeked into the world of decision-making in Chapter 17, with serial and layered behaviors supported by pattern matching to trigger changes. The gray area of decision-making fills the gap between machine control and raw sensory data. You have inputs, you have outputs—how do you tie the two together?

For industrial machines, the senses are used to fine-tune the control. There is no sensing and reacting to the environment in a big sense, no strategic decisions based on external input. The sensor is used to make the industrial robot more accurate, to keep the process within defined limits. All of the important decisions were made by the human who wrote the program or script that is driving the robot.

As we approach the intelligent robot, the subject of how it chooses its behavior grows larger. Many of the mobile robots don't make decisions, but operate on simple reflex. Bump into something, avoid it. Receive an instruction and follow it. Sense a low battery and return to home. Instinct, like a bug or worm, reacting in a basic way to its environment. Surprisingly, even that much intelligence has taken researchers decades and several keen insights to develop. At this rate, the lofty goal of mimicking human intelligence seems impossible.

It can be interesting to introspect, to look inside our own mind and try to see how we make decisions. Looking at our intelligence from the inside provides special difficulties since we are limited in our ability to sense our own thought processes. The part of our mind that is doing the thinking must have some additional mechanism behind it that we cannot see—it is behind the scenes, running the system.

Trying to work out the thought process from the outside has its own problems. We can analyze and hypothesize endlessly, design and execute experiments, but we only see the final results of the opaque black-box that is our subject. We can't see the inner workings. New technology is helping to pry open this box, but it still limits us to seeing which corners of the brain are active during what activities. Like watching a machine and noting which gears are spinning during any given task. Illuminating in its own way, but still a long way from explaining how and why.

At the other extreme, we have been taking the individual neurons that make up the brain and seeing what makes them tick. As early as the 1950s, with such work as A.L. Hodgkin and A.F. Huxley's analysis of the giant squid axon, we have had detailed models of how a neuron works. This bottom-up approach tries to determine how the larger structures of the brain behave using systems of these neurons. The hope is that the top-down approaches will eventually meet the bottom-up approaches somewhere in the middle, giving us a comprehensive model of the brain.

There is also some hope that the brain is composed of many modules or layers, each of which is constructed similarly even though they may be applied to different tasks. Nature has a way of reusing designs.

Simple reflexive control is a tight control loop between input and output. Touch a hot stove and your finger registers immediate pain, and a neural reflex snatches your hand from the stove. The next layer up can take the pattern of sensory input and abstract it into the pattern for "burning hot."

There will have been a stream of sensory input before and after the burning. The eyes receive streams of information, patterns of dots that are abstracted into coherent shapes, which will in turn be constructed into familiar shapes, ultimately receiving labels like "stove" and "red burner." Another layer can lay this sequence flat and see the relationships over time—the "red burner" combined with the motions that brought the hand into contact with it, and then the hard-wired pain. The next time "red burner" registers itself, a layer of intelligence can run forward in time, remembering or simulating experiences that relate to the burner. It can then avoid actions that repeat the previous pain.

Reflexive actions respond to stimulus now. More complicated reactions look into the future and choose actions that lead to a desired outcome. It can make decisions. Hunting creatures, especially, need to predict the future. Where will the prey be in five seconds? How will this animal react to noise, to smell, to motion? The better it is at predicting the future, the longer ahead it can plan. When the plan goes farther into the future than the prey's ability to look ahead, the hunter can affect a capture. There is still something to be said for brute force, but why work hard when you can think ahead instead?

Another aspect of this ability to predict the future is the ability to model another creature's thought process. The predicting brain must not only record and be able to play back its own experience, but be able to act as a model of another brain to be able to predict its actions.

You could argue that decision-making is a reflex of pattern matching coupled to specific actions. The pattern matching is not tied to the here and now, but can match patterns predicted into the future. The hard-wired rewards and punishments provided by our bodies provide value determinations for these actions, enabling decision-making. It is highly unlikely that the brain, human or otherwise, is this simple.

Mapping

Moving beyond philosophical concerns, we return to the subject of learning.

A mobile robot in an unknown environment has, as a primary concern, the need to develop a model of its environment. A map. Given a simplistic room like the one in Fig. 18-1, you would think it would be easy for a robot to learn the layout.

One first step is to try to tame the infinite spaces of this room. The robot has limited memory and we need to find a way to store the layout of this room in it. One way is to lay a grid over the room (Fig. 18-2) and then record information in each grid square as we find it. This grid layout mimics the memory structure of the computer and is easy to represent. As the robot wanders around the room, it notes what it finds in each grid square as it passes through it (Fig. 18-3).

The robot has moved a long distance and it hasn't learned much about the room. What if we added the ability to see out to the sides of the robot? There are optical sensors that can provide a type of visual bumper, returning

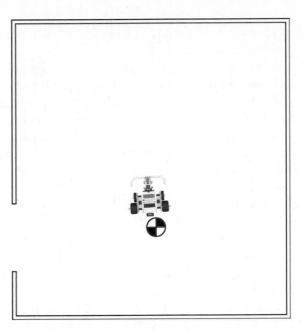

Fig. 18-1. Simplistic room.

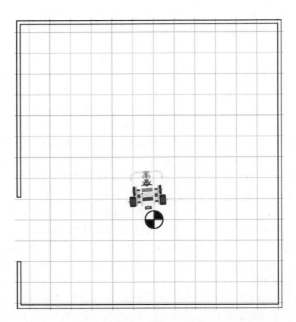

Fig. 18-2. Grid overlay.

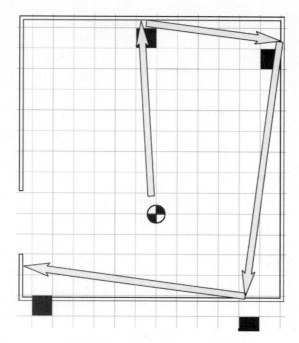

Fig. 18-3. Bumper hits in the room.

the distance to an obstacle. Assuming we can see out to the sides of the robot by three squares, this same path may return more information, as shown in Fig. 18-4.

ODOMETRY

Mapping assumes that the robot has some way of knowing where it is as it travels. The simplest technique for this is *dead reckoning*, technically known as *odometry*, the measurement of distances traveled.

The math behind odometry is surprisingly simple. It requires an accurate measurement of the distance each wheel has traveled. From this, it can calculate the change in position and rotation of the vehicle.

Figure 18-5 illustrates the odometry math. The distance traveled by the left wheel is D_L, and by the right wheel D_R. The width of the robot, from where the left wheel touches the ground to where the right wheel touches the ground, is W.

At time $T = 0$ the robot is facing in the direction O_T, which is specified in radians. If the robot turns during its travels, it ends up in a different

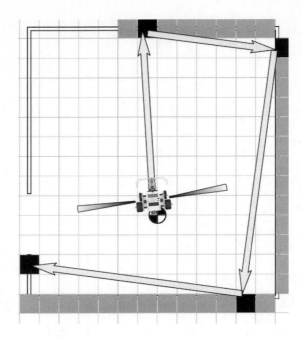

Fig. 18-4. Side sensors in the room.

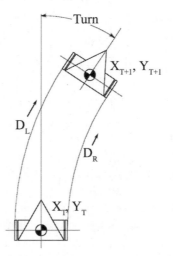

Fig. 18-5. Idealized robot for odometry.

orientation O_{T+1}. This new orientation is calculated as a ratio of the wheel's relative motions with the robot's width:

$$O_{T+1} = O_T + \frac{(D_R - D_L)}{W} \qquad (18\text{-}1)$$

The straight-line distance the robot traveled is the average of the two wheels' travel distance:

$$D_T = \frac{(D_R + D_L)}{2} \tag{18-2}$$

The new position of the robot in the X, Y grid is then approximated by:

$$X_{T+1} = X_T + D_T \times \cos(O_{T+1})$$
$$Y_{T+1} = Y_T + D_T \times \sin(O_{T+1}) \tag{18-3}$$

The catch is that this motion is only a straight-line approximation of the actual curved path.

ODOMETRY ERRORS

Even if we used better, more complicated, math there would be errors recording the motion. The wheels don't contact the floor in a precise mathematical point so there is some error in W. If there is any slipping or shifting of the wheels, the measured distances do not match the actual distance traveled. Also, the sensor used to measure the wheel motion is not infinitely precise, which adds some uncertainty into each measurement. All of this adds up to errors, especially errors in calculating the robot's turns.

There are also errors in the sensors, so they don't report the exact truth of the situation. Even if we "blur" the information returned from the sensors, the odometry errors can still put us out of the ballpark. Figure 18-6 represents this.

This is a tricky diagram. The wide path is where the robot thinks it is going. The narrow path illustrates the actual path through the environment. The blurred sensor marks show how the robot's sense of the environment varies from the actual environment—it is distorted and inaccurate. The more the robot travels, the more distorted its map becomes, until it's a useless blur. Researchers who use odometry spend a bunch of their time correcting for these built-in errors. One of the lessons to learned from this is that robotic tasks are never as simple as they seem at first glance.

Supervised Learning

The example of learning a map, above, is an instance of throwing the robot into an environment and expecting it to learn without any supervision. It is easier on the AI, if not for you, to provide a bit more direction.

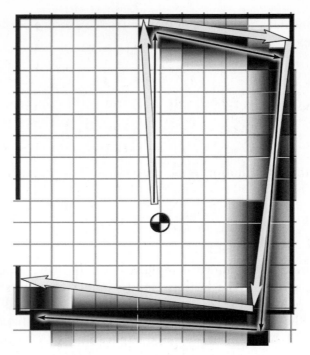

Fig. 18-6. Sensed room versus actual room.

Supervision can be applied to the mapping problem. The robot is given a map that is already filled in and it tries to keep track of its position on this map. The odometry methods place the robot on the map as it wanders across the center. Then, when the robot senses a wall it can compare its sensory input to the map and adjust its concept of where it is on the map. This combines the data from the odometry system with sensory data and applies it against the template of the built-in map. If the robot reaches a corner, it can perform an even more accurate realignment of its internal map, since the corner constrains the robot's position along two axes of motion.

Many artificial intelligence technologies use a form of *supervised learning*. Supervised learning provides both the question and an answer, and the system learns the connection between them. We apply a training pattern to pattern-matching neurons in Fig. 18-7.

In this type of supervised training, the programmer sends an input pattern x_n to the neurons as well as the desired output T_m, or training pattern, upstream to the outputs y_m. The input is processed and the actual results are compared to the desired output. The difference, or error, is propagated

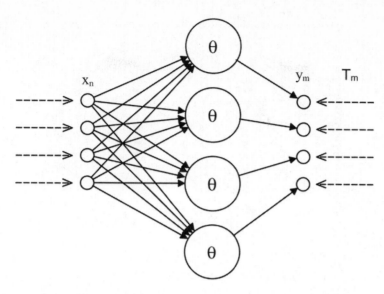

Fig. 18-7.   Supervised training of pattern-matching neurons.

backward through the neurons. This error propagation changes the values in the neuron's templates so they are more likely to generate the desired output for this input.

Over time, and with a reasonable set of training patterns, each neuron in the network will come to recognize a different distinct input pattern. Inputs that are near, but not quite the same, as a known pattern will activate a neuron with a similar pattern more than a neuron with a distant pattern. This simple recognition system is known as a *perceptron*, and it was one of the first neural network models.

This one-layer network is a good start, but it is limited in how well it can abstract information. For example, it was proven to be unable to learn the XOR function. This limitation was so crippling that neural networks were abandoned as a technology for a long time, until some new insights were developed.

By adding another layer, the network can learn complex associations between input and output (Fig. 18-8). This multilayer perceptron (MLP) is a truly useful processing module. Training is still supervised, and the neural template adjustments are still done automatically.

Where information in our map was stored in a regular grid, information in these networks is stored in the template in each neuron. In Fig. 18-7, these templates even made sense. Moving into Fig. 18-8, the values matched by the neurons are less straightforward. One of the drawbacks to multilayer neural

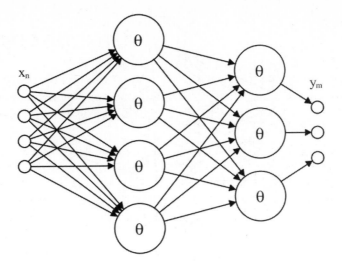

Fig. 18-8. Multilayer perceptron.

networks is that you can't really pry the top off of the system and see what it is doing. The numbers the network learns don't necessarily make sense by themselves, but work together to create a useful answer. In a sense, each piece of information is stored across the entire network. Some authors have even compared the brain's method of storage to a hologram.

SUPERVISED ROBOTIC LEARNING

Assuming you have a working neural network, how does this apply to your robot?

The input pattern could be values from a number of sensors on the robot. The output is then the control signals that drive the robot. This is a reflexive type of intelligence that can be used for low-level control. You can train this type of system to perform simple tasks. Perhaps the easiest way to train it is to show it how to perform the task.

The trainable robot is a combination of self-guided machine and telerobot. In training mode, a human operator controls the robot's actions, guiding it through the task and around the environment. The neural network "watches" as the operator guides the robot. It matches the sensory inputs with the control signals sent by the human operator. When turned loose on its own, the sensory signals stimulate the neural net, which generates control signals that are similar to the human's control during training.

Unsupervised Learning

Supervised learning has its place, and it provides great utility for many different pattern-matching problems. There are times, however, where you want your system to learn on its own. Not all problems are easy to break down into supervised learning sets.

Fortunately for us, almost all "real" phenomena occurs as a continuous stream of events, each of which is related, by physics and the flow of time, to the previous one. Our sensory input is continuous, with strong internal structures and a definite sense of cause and effect. Neural networks can make use of this coherence to create statistical groups of sensory patterns. These groups divide the network's experience into significant bundles using *unsupervised learning*.

Some unsupervised networks, called *self-organizing maps* (SOM), learn by comparing the current sensory input against the history of all inputs as embodied by the patterns of their neurons. Where supervised learning is given the answers, unsupervised learning discovers the "answers" in the very textures of its experience. The internal details of the neurons are still the same: computational neurons containing templates that are matched to the input pattern, to create an output activation. But in training, the activated neurons are not compared to the desired output but instead to their neighbors. In a neighborhood group, the best matching neuron adjusts to better match the input, while more distant neurons move away from that pattern.

Once trained, each neuron in the map represents a point in state space. *State space* is like a map of all possible inputs, in all possible combinations. In reality, only a small set of input combinations are going to appear on the input of the network. The network adjusts to represent the inputs that appear and pretty much ignores areas of state space that are not relevant.

Mapping the self-organizing map to robot control is a bit tricky. In theory, the system could learn what actions are appropriate to which inputs through trial and error, and with some form of reward and punishment system. There is a system of learning called *reinforcement learning* that does just this.

In practice, the SOM may be hooked up to a more traditional multilayer perceptron. The activation levels of the neurons in the map become inputs to the MLP, which is trained so that the correct control signals are generated. In this case the SOM is a pre-processor for the MLP. It can take a large, poorly defined, and imprecise set of inputs and simplify them.

Swarm Robots

The robots you see in the movies, the C3POs and the androids, are built in our own image. A single machine with flexible manipulators (arms, hands, and fingers) capable of traversing most forms of terrain (legs, feet) and with a single, complex control system (brain).

Industrial robots are almost the direct opposite of this. They are special-purpose machines specialized for a single task, they tend to be fixed in position, and their control system is little more than a script reader playing back preset instructions.

There is a third approach to robots, almost alien in nature. Their inspiration is in the complex communities and behaviors of ants and bees. A single ant is not very complex. Building an ant brain is a far more approachable task than building a human brain. Ants in groups have more complex behavior than an ant by itself. Much like a brain has behavior beyond that which could be anticipated by looking at a single neuron.

This thinking brings us to swarm robotics. Each robot is a small, inexpensive, and simply programmed unit. It has the ability to communicate with neighboring robots and it is deployed with a large number of its siblings. Any small subset of these robots can be destroyed or lost and the swarm can still fulfill its job. The robots are inexpensive so there can be a bunch of them, and since there are a bunch of them any one of them is expendable.

This configuration of robots, the swarm, lends itself to exploration. Exploration of space leaps to mind, but there are also important exploration tasks on Earth, such as swarming over a disaster area looking for survivors. Swarms will also be used in wartime, to scout out the urban landscape and report on conditions. Swarms of flying robots, another topic that has generated a lot of interest lately, are especially useful for scouting.

Related to swarm robotics is modular robotics. In this case, the swarm comes together, the individual robots attach to each other physically. This makes them into a larger robot for the purpose of completing a task that is not possible for a single, smaller robot.

Agents

As the world goes online, many of our activities are no longer in the realm of the physical but involve traversing and manipulating our new electronic realities. There will be online tasks that we don't want to do ourselves because they are hard, tedious, or otherwise better left to automation. And some of these tasks will take a fair level of intelligence.

What is needed is an electronic "robot" with many of the traits of its physical counterpart. The ability to travel around its environment, in this case the Internet. The ability to sense and manipulate this environment. And the ability to match incomplete or unclear patterns and to make action decisions based on them.

Such a virtual robot is known as an *agent*. Agents tend to operate in swarms, interacting with other agents with perhaps different capabilities. Agents lack the physical dimension, so there is none of that pesky, expensive hardware to worry about. They are robots in their most abstract form, pure information.

People have been creating simulated robots in simulated environments almost as long as we have been building computer-driven robots. It is easier and faster to perform experiments in the computer before committing the design to solid form. But those are still just simulations of a robot whose ultimate operation is to be in the physical world. An agent's environment *is* the world of information.

Summary

This chapter gave an overview of some artificial intelligence concerns, and how AI techniques could be applied to robots.

First we stepped back and looked at the role of pattern matching in the decision process. Then we saw how the apparently simple task of mapping a room is complicated by the realities of mechanical construction and inaccurate sensors. The neural systems we explored in Chapter 17 were expanded to allow them to be trained and then to train themselves. In closing, we mentioned some new, nontraditional robotic systems such as swarms of robots and online agents.

Quiz

1. What are the sensory inputs on industrial robots used for?
2. Name and describe a technique used to calculate a robot's position.
3. What is the problem with the dead-reckoning system from question 2?
4. What are the two types of neural network learning?
5. Do robots always have to have a physical body?

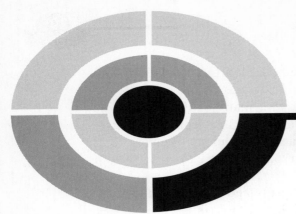

Chapter 2

1. 1.2 meters (4 feet). In the equation $d = (a \times t^2)/2$, a is the acceleration from gravity which is $9.8\,\text{m/s}^2$. Time t is 0.5.

$$d = \frac{9.8 \times 0.5^2}{2}$$

$$d = \frac{9.8 \times 0.25}{2}$$

$$d = 1.2$$

Bonus question: First, you need to turn the equation inside out, so solve for t:

$$t = \sqrt{\frac{2 \times d}{a}}$$

So, solving for t with $d = 3$ and $a = 9.8$ gives a travel time of 0.78 seconds:

$$t = \sqrt{\frac{2 \times 3}{9.8}}$$

$$t = \sqrt{0.612}$$

$$t = 0.78$$

2. This is a tricky question, since the details are spread around. The force is the mass times the acceleration, sure. But what is the acceleration of a ball being stopped in a tenth of a second?

Acceleration is the change in velocity over time. We know the ball was moving at 10 meters in 1 second, and once it hits the wall it is moving at 0 m/s. So the change in velocity is −10 m/s. We can ignore the negative sign, it is just telling us that this is a deceleration instead of an acceleration.

Since this change in velocity accurs across 0.10 seconds, we see 10 m/s divided by 0.1 seconds, for an acceleration of 100 m/s². So the applied force is 50 newtons:

$$F = 0.5 \times \frac{10}{1 \times 0.1}$$

$$F = 0.5 \times 100$$

$$F = 50$$

A soft wall changes the velocity to zero across 0.25 seconds, for an acceleration of 40 m/s. The soft wall only needs to apply 20 newtons of force, but it does it across a longer time span:

$$F = 0.5 \times \frac{10}{0.25}$$

$$F = 0.5 \times 40$$

$$F = 20$$

3   0.5 kg at 10 m/s has a kinetic energy of 25 joules. This is a straigtforward solution of the kinetic energy formula:

$$KE = \frac{0.5}{2} \times 10^2$$

$$KE = 0.25 \times 100$$

$$KE = 25$$

At 20 m/s, the kinetic energy is quite a bit larger:

$$KE = \frac{0.5}{2} \times 20^2$$

$$KE = 0.25 \times 400$$

$$KE = 100$$

4.   Another easy solution, using the potential energy equation:

$$PE = 0.5 \times 9.8 \times 4$$

$$PE = 19.6$$

Now, you know that the ball dropped from 4 meters, so the time it would take to hit the ground is 0.9 seconds. Since it crossed these 4 meters in 0.9 seconds, you might think that its velocity is 0.9 m/s,

giving a kinetic energy of less than 5 joules. Where did the other 14 joules go?

The problem is that this velocity is the average velocity across the entire fall, most of which is spent going really slow. The kinetic energy would need to be calculated for the final velocity of the ball when it hit. This is trickier, and is not explored here. Sorry.

Chapter 3

1. Statics is the study of the forces on objects as they just sit there. Dynamics is more concerned with forces of objects in motion. Note that kinematics is also the study of motion. They go hand in hand: kinematics of motion, dynamics of the forces behind the motion.
2. Cutting paper uses shearing forces, and scissors are often called shears. Crushing a grape uses compression force and breaking a string involves tension. Examinations also include tension, but of a different sort.
3. The vertical part of the motion is the part that fights gravity. This is calculated as the sine of the angle:

$$v = \sin(10)$$

$$v = 0.17$$

So as we push the robot up the ramp, it is fighting 17% of gravity's force, or an acceleration of $1.67\,\text{m/s}^2$. Now that you know the acceleration fighting the climb of the robot, what else can you calculate, assuming you also know the weight of the robot?

4. Your knee, like your elbow, is the fulcrum for a second-class lever. Your ankle, however, is the fulcrum in the center of the first-class lever that is your foot.
5. 25 teeth, each taking 0.125 inch, uses up 3.125 inches total. This would be the circumference of the gear, so the gear's radius is just about a half inch:

$$c = 2 \times \pi \times r$$

$$3.125 = 6.283 \times r$$

$$r = \frac{3.125}{6.283}$$

$$r = 0.497$$

The big gear has a diameter of 4.97 inches giving a radius of 2.485 inches. The circumference is then:

$$c = 2 \times \pi \times 2.485$$
$$c = 15.61$$

Leaving room for 125 teeth. Okay, 124.9 teeth. The math, as shown, rounds the numbers, losing a bit of accuracy.

The mechanical advantage is then 5:1 or 1:5, depending on how you look at it. Note that this is essentially the same as 15.61:3.125, the relative circumferences of the gears. For that matter, it's also the same as the ratio of the diameters, 0.994:4.97.

Chapter 4

1. The electrons in a conductor are free to move around, while they are stuck in place in an insulator. Both conductors and insulators are filled to the brim with electrons, it's just a matter of how easy they are to shift around.

2. This is a trick question. There is no one good answer! Electric charge is held by the charged particles, the electron (negative) and proton (positive). Voltage is the difference in electric charge from one place to another. Current is the flow, or movement, of electric charge.

   When you "make electricity" you are not creating electrons. You can, however, move them around (current) to create differences in charge (voltage).

3. A changing magnetic field creates an electric field. A changing electric field creates a magnetic field. Together, these fields create a self-sustaining wave that radiates out through space. When this wave intersects a metal antenna, it moves the electrons in it a little bit, creating a tiny signal that can be amplified and listened to.

   For an electric generator, a continuously moving magnetic field creates an electric field. This moves the electrons, creating electrical current (see Question 2).

   For an electric motor, the electrons are moved by an outside charge pump (battery or generator), causing their wires to move against a background magnetic field.

4. A circuit is a track that electrons run through. The simplest circuit is a wire connecting two ends of a battery. This is the Indy-500 of circuits,

a simple unbroken circle (don't do this, it's bad for the battery). We normally drop additional components into this track, creating something closer to an obstacle course or, for some poorly designed circuits, a demolition derby.

Chapter 5

1. A schematic. You can actually have schematic diagrams for all kinds of things, but they tend to be called by different names, such as blueprints, or those posters you sometimes see in health class.
2. "Gnd" is ground, which is normally the negative terminal on your battery or generator. "R1" is the first resistor in your schematic. "C5" would be the fifth capacitor, and "S2" the second switch.
3. I personally like to build my temporary, or prototype, circuits on a breadboard, but there are other ways of creating prototypes as well. Once I'm happy with a circuit, I like to put the permanent version on a printed circuit board, or PCB. This could be a custom-built one, or one that mimics the layout of the breadboard.
4. A resistor opposes the current in your circuit. As do the wires in the circuit, and everything else, but a resistor is the specific component I was looking for here.
5. Georg Ohm, though you only really need to remember his last name. Ohm's Law is:

$$V = I \times R$$

$$I = \frac{V}{R}$$

$$R = \frac{V}{I}$$

6. We know the voltage (9) and the current (0.015), so we need to calculate the resistance:

$$R = \frac{9}{0.015}$$

$$R = 600$$

A 600 ohm resistor will work, though one slightly smaller or bigger is also fine. LEDs aren't too picky.

A single 1.1 kΩ resistor would pass a stingy 8 mA through, making for a dim LED. However, two of these in parallel give 550 Ω, for a shiny 16 mA current for our LED.

7. Good! You didn't look at Fig. 5-19, did you? Well, do so now and compare your diagram to it.

A Wheatstone Bridge is a pair of voltage dividers that can be used to isolate small changes in a resistive sensor, or to determine the value of an unknown resistance.

8. Robots, as we understand them today, use electronics in all aspects of their operation. Their brains are normally electronic computers and, when they aren't full computers, their brains are complex electronic circuits. The interface from sensors to the brain is a layer of electronics, as is the interface from the brain to the motors that drive the robot. If you are going to understand and build robots, it helps to be comfortable with electronics.

Chapter 6

1. Its behavior.
2. The old automata used open-loop control systems. You could also say they used gears, cams, and so forth, but these are the pieces of the open-loop control.
3. Closed-loop control gives the controller feedback about the object under control. Feedback allows the controller to adjust to unusual circumstances.

Chapter 7

1. We listed SPST, SPDT, DPST, and DPDT switches. Single or Double Pole by Single or Double Throw. The single-throw varieties also come in normally open and normally closed forms.
2. No, not open-loop control, that was the last chapter. In this chapter we talked about how cams can work together to create a control system.
3. Both punched cards and punched tape were both used to program computers and automatic looms.

4. These lights spell out 10110011 in binary. Taken in two parts, 1011 is an 11 in decimal, or B in hexadecimal. 0011 is 3 no matter how you slice it. All together, however, it converts to the value of:

$$\left(1 \times 2^7\right) + \left(0 \times 2^6\right) + \left(1 \times 2^5\right) + \left(1 \times 2^4\right) + \left(0 \times 2^3\right)$$
$$+ \left(0 \times 2^2\right) + \left(1 \times 2^1\right) + \left(1 \times 2^0\right)$$
$$= 128 + 32 + 16 + 2 + 1$$
$$= 179$$

5. There are many right answers here, as there are many types of computer languages. Most computer systems consist of a set of instructions (the program), a bunch of slots that can hold numbers (memory), and a circuit in the middle that interprets the instructions and manipulates the memory (the central processing unit, or CPU).

Most such systems include commands to get a number from a position in memory. When a number has been retrieved, there are usually instructions that can perform math on that number and the number in the CPU'S other hand (some CPUs have two hands to hold numbers, others might have sixteen or more). The CPU can also compare two numbers and choose which instruction to run next based on the result.

Chapter 8

1. Your car's wheel uses a rotating joint. There are many ways to reduce the friction in a rotating joint, including bushings and ball bearings.
2. Your kitchen drawers slide along a track. Adding some slippery grease to the track will make it run smoother. However, if you take a ball bearing and straighten it out to make a ball-bearing track, you remove even more friction. Most joints, in fact, work better rolling on balls than rubbing against each other.
3. Other than despair, or reattach your motor, you can connect them with a shaft that includes one or more universal joints. This is a sneaky lead-in to Chapter 9, Power Transmission.

Chapter 9

1. An idler can be used to tension a chain or keep it from tangling with itself. This also works for belts and other flexible loops.
2. A chain can carry power (torque) from one place to another. The chain itself can have special attachments that let it act as a tank tread, part carrier, or some other flexible, moving device.
3. Gears can carry torque from one place to another, including around corners. They can also increase torque (while decreasing speed) or increase speed (while decreasing torque). Gears can reverse the direction of rotation.
4. Split gears are used to reduce backlash. Backlash is the "slop" between gears, and it introduces inaccuracy in the gear train. Split gears help improve the accuracy of a gear train, especially if it rotates both forward and backward during its normal operation.
5. Spur gears are the basic flat gears, and are used to transfer and manipulate torque in a machine.

   A crown gear has teeth poking out from the face of the gear. It meshes with spur gears to transfer rotation around a 90° corner.

   Bevel gears mesh with other bevel gears to transfer rotation around corners. While they normally sit at a 90° angle from each other, different angles are also possible.

   Worm gears mesh with spur gears. These gears reduce back-driving, provide a large speed reduction and torque increase, and transfer torque 90° along a different axis than bevel and crown gears.
6. Couplers normally adjust for slight misalignments between the power source and its destination. Fancier couplers can do more than this, and in those cases they are called transmissions.
7. The differential.

Chapter 10

1. You plug your toaster into alternating current (AC). When you lick a 9-volt battery, your tongue is tingling to the tune of direct current (DC).
2. The frequency of a signal is the number of times it repeats its pattern in one second.

3. A diode conducts electric current in one direction but not the other. By putting it in the circuit backwards, it will not conduct. If you crank up the voltage it will eventually fail and let the magic smoke out. When you turn up the reverse voltage on a Zener diode, it will reach a point where it does conduct. Of course, if you turn the voltage up too high the Zener gets hot and melts its important bits just like any other component.

4. A capacitor consists of a dielectric, or insulating layer, and a thin electrode on each side of the dielectric. Of course, you also need terminals to be able to plug the capacitor's electrodes into your circuit.

5. A capacitor does not conduct current, but power can flow through it. A changing electric charge in one of the capacitor's electrodes reaches through the dielectric to wiggle the electrons in the other electrode. This wiggling can do work, but there is no flow of electrons through the capacitor.

6. A resistor and a capacitor can join to make different types of AC filters. While there are several answers to the second part of this question, the first equation used for RC circuits describes the time constant $\tau = R \times C$.

Chapter 11

1. Current in a wire creates an electromagnet. And an inductor. The second coil turns it into a transformer, which is like a gear train for electric charge. Other uses for electromagnetism include the relay, motors of many kinds, and generators.

2. Inductors store a magnetic field. Like a flywheel, they resist change, so AC signals will be smoothed out by the inductor. DC, however, passes through with nothing more than a bit of resistance.

3. Resistors and inductors can be combined into RL filters, though it is more common to use resistors, capacitors, and inductors together in filters.

Chapter 12

1. There are passive and active components. Semiconductor-based components are active while resistors, capacitors, and inductors are considered to be passive.

2. Hopefully you remember some of it! Free electrons, holes, majority and minority carriers, dopants, depletion zones. There is a lot of fancy wordsmithing needed to talk about semiconductors. Even more important than the words, which you learn with repeated use, are the concepts and how electrons behave inside the semiconductor lattice.

3. Doped semiconductors comes in two flavors, n-type and p-type. An n-type semiconductor has an excess of free electrons. A p-type semiconductor has the reverse condition, an excess of electron-binding holes in the lattice. Both free electrons and holes are charge carriers, and can move freely through the semiconductor.

4. The depletion zone is the junction between n-type and p-type semiconductor chips, where the holes and free electrons join into an insulating layer. Which insulates. A forward-biased depletion zone grows smaller until it conducts current. A reverse-biased depletion zone gets bigger and continues to refuse to conduct current.

5. A transistor is a voltage-controlled insulator. It uses a small forward bias across one of its depletion zones (the emitter/base current) to control the larger conductance through the entire transistor (the emitter/collector current). This acts like an amplifier, taking a small signal on the base and turning it into a large signal at the collector.

6. The answer is in Fig. 12-18. I bet your picture looks just like it.

Chapter 13

1. The syntax consists of the symbols in the language. Semantics describes what they mean, their purpose in the program.

2. Subroutines are collections of computer statements that, together, perform some task. Subroutines make it easy to organize your program into functional blocks, as well as to use and reuse those blocks as needed.

3. Watching an input means to poll the input. Of course, if the input interrupts the program, it is an *interrupt*.

Chapter 14

1. You have a lever! Levers can be used to make you stronger. Or you can push on the short side to make a quick action using a short input.

A lever with the fulcrum in its center doesn't give you strength or speed, but does reverse the direction of motion.

2. The wheel could be your power source, in which case the sticks can be used to create straight-line motion from it. If you have a steam cylinder, the two sticks can work backwards, turning the linear motion into the rotation of the wheel.

3. A four-bar linkage can create parallel motion. Or, with proper design, it can move its coupler between two predefined positions.

4. You can limit rotation by attaching a ratchet gear to the wheel and using a spring-loaded pawl to prevent reverse rotation.

Chapter 15

1. The human operator makes all of the high-level decisions. They are the intelligence behind the curtain, while the robot gets to worry about the low-level reflexive actions. For really lame systems, the operator may be in direct control of the motors.

2. A semi-autonomous robot is only partially under operator control. The operator makes the big decisions and then turns the robot loose to implement them. Many times, it is impossible to have direct control of the robot, so it needs onboard intelligence to control itself between commands. Even if you have mostly direct control, wireless communication can be disrupted so the robot still needs to behave reasonably until the connection can be reestablished.

Chapter 16

1. Communication is a way to transmit information between two entities, such as people. It can be in the form of language, gestures, tone of voice, facial expressions, and so on.

2. A program is a series of numbers to the computer. The arrangement of the electronic components that make up the computer's processing unit provides meaning to the numbers.

3. The Turing machine is a generic, universal computer. It consists of a long tape covered with symbols that act as data for the machine. A read/write head manipulates the tape, while an internal state register

keeps track of the machine's status. A transition table provides guidance to the machine.

4. There are different ways to break down the world of computer languages. One arrangement consists of control flow, data flow, and declarative (no flow) languages. Of course, there is also human language, which doesn't fit any single category.

Chapter 17

1. I couldn't define intelligence either, not with any great level of success. For most robots, there is no need for any type of intelligence. Their needs are limited to the rote execution of instructions and some low-level reflexes to shut down when something goes wrong.

2. PID (proportional, integral, derivative) controls are used to drive all sorts of systems, from motors to elevators (which are mostly just a box and a motor), to factory systems. They adjust their output based on an error measurement, the accumulation of error measurements, and even a prediction of error measurements to come.

3. The tricky aspect of sequential behaviors is often the decision as to when to change the behavior. For many cases this is easy—punch three holes and then change tools. In other cases it is hard, such as "find Fred and give him this message." The driving around is not too hard, and replaying a message is easy, but how do you know it is Fred?

   Layered behaviors have the same pattern-matching situation. How does one layer know when to subsume the next?

   Of course, in many cases the behaviors are hard to program as well. In those cases, it's all hard.

4. There are probably others, but we discussed neural networks. In the simple one-layer network explored in this chapter, each neuron compares its template, or pattern example, with the signals it is receiving. The one with the best match shouts its result the loudest and wins the vote.

Chapter 18

1. Industrial sensors provide feedback to the controller so it can fine-tune the position of the robot's actuators.

2. If you know the motion of a robot's wheels, you can calculate its rough position using odometry. The distance the center of the robot travels is the average of the distance its two wheels traveled. The change in the robot's angle is the difference in the wheel's motions divided by the width of the robot.

3. Odometry suffers from the accumulation of errors, so the position and orientation information slowly becomes useless.

4. Neural networks can be trained with supervised learning, where situations and answers are presented together, and by unsupervised learning, where the network figures out the structure of its universe by exploring it.

5. Agents are like robots but without that cumbersome physical component.

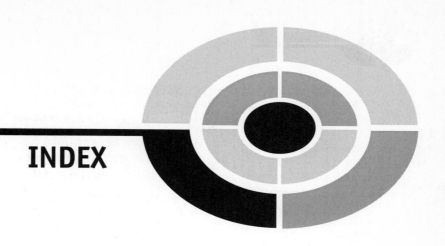

INDEX

ABOUT THE AUTHOR

Edwin Wise is a software engineer with twenty-five years of experience. His expertise and interests range from electronics and microcontrollers to software development, AI, and robotics. He is the author of McGraw-Hill's *Hands-On AI with Java* as well as *Applied Robotics*, *Applied Robotics II*, and *Animatronics: A Guide to Animated Holiday Displays*, published by Delmar Learning. He lives in Austin, Texas. He can be reached at edwin@simreal.com, unless you are trying to sell him something.